T0393294

Health and Healthcare Policy in Italy since 1861

Francesco Taroni

Health and Healthcare Policy in Italy since 1861

A Comparative Approach

palgrave
macmillan

Francesco Taroni
Department of Medical and Surgical Sciences
University of Bologna
Bologna, Italy

ISBN 978-3-030-88730-8 ISBN 978-3-030-88731-5 (eBook)
https://doi.org/10.1007/978-3-030-88731-5

This Palgrave Macmillan imprint is published by the registered company Springer Nature Switzerland AG.
The registered company address is: Gewerbestrasse 11, 6330 Cham, Switzerland

PREFACE

The book provides a historical account of health and healthcare in Italy since 1861, the year of its unification, to the present COVID-19 pandemic. This aims to fill a void in the English language literature about health, medicine, and healthcare organization in Italy for students, teachers, and scholars of Italian politics, history, and culture. The number of important books about the political, economic, and cultural history of Italy pays only slight attention to health and health policy. This void is an important gap since health and healthcare policy and politics are among the most distinctive characteristics of a country.

The periodization of the book places the history of health, healthcare, and their institutions firmly in the political and social context in which they developed and evolved, corresponding to the classic division of Italy's political history into a Liberal period, Fascism, and the Republic. A distinct health system model characterizes each political period, although significant continuity is also apparent. In the liberal state, health was legislated by the state but funded and administrated by *comuni* (municipalities). Domiciliary care was provided by *medici condotti* (municipal doctors) and residential care by a maze of private, mostly religious, charities (*Opere pie*, or Pious Organizations). In the interwar years, the fascist regime introduced an occupationally based social insurance system run by autonomous institutions, *casse mutue* (sickness funds), reproducing its corporatist polity. This model was maintained in the first thirty years of the Republic. In fact, while the fascist regime adapted health legislation and institutions from the liberal era, many of its institutions became the foundations of the republican welfare state. The establishment in 1978 of the *Servizio sanitario nazionale* (SSN—National Health Service) is a watershed in the organization of health in Italy. The SSN is the first continental system of health and healthcare organization based on Beveridge's principles of universalism, equity, and solidarity, financed by general taxation and modelled on the British National Health Service (NHS). The SSN transformed individual entitlements based on occupation into equal citizenship's rights based on health needs and put health politics and policy on a different path from the rest of the Italian welfare system. Health policies and institutions will also be considered by examining similarities and differences in addressing "exemplary" conditions, including pellagra (a dietary deficiency disease mostly affecting the poorest peasants) and three infectious diseases, cholera, malaria, and tuberculosis. Malaria, for example, was addressed in the liberal period first with the traditional programmes of *bonifica* (land reclamation) in use since Roman times and then by a science-based model of disease-specific, vertical intervention through human *chininizzazione integrale* (total quininization). This contrasts with the fascist "totalitarian" policy of *bonifica integrale* (whole land reclamation), comprehensive of hydraulic, agricultural, and human interventions, and is more similar to the strategy of environmental sanitation by DDT adopted in the campaigns in the aftermath of the Second World War.

Some readers will surely complain about the cursory coverage of certain essential topics and criticize the lack of discussion of some major issues, such as medical and nursing education, the diffusion of critical medical technologies, and important diseases such as polio or conditions such as obesity, to name just a few deficiencies of the book. One hundred and fifty years is a very long time span to cover in the rapidly evolving domains of health, medicine, and health organization and policy. Selecting crucial events is necessarily idiosyncratic.

A preface is also a place to acknowledge debts. I owe a personal and academic debt of gratitude to Francesco A. Manzoli, Elio Guzzanti, and Joseph S. Gonnella for the benevolence and wisdom which they have generously bestowed on me at different moments of my professional life, in Rome at the Istituto Superiore di Sanità and Ministero della Sanità, and at Thomas Jefferson University's Sidney Kimmel Medical College in Philadelphia.

From Maria Malatesta's vast culture and rigorous historical method, I have tried to take inspiration on how to think as an historian. From his retreat on the top of the mountains, Marco Biocca has kindly provided valuable comments on several chapters. Over the years, George France has always shared with me his deep knowledge, bright mind, and Scottish wit. Several chapters of this book, and particularly Chaps. 8 and 9, build on his knowledge and the work we did together. This book would have been much better if he had the chance to participate in its writing. Daniel Z. Louis, a distinguished scholar in health service research, managing director of the Sidney Kimmel Medical College's Asano-Gonnella Center for Research in Medical Education and Health Care, and particularly, *amico di una vita*, has provided outstanding help in preparing the manuscript and has made observations and comments on each line of the book with competence, patience, and friendship. Maria Luisa Moro, director of the Agenzia Socio-Sanitaria Regionale of the Regione Emilia-Romagna and my wife, has always been patient enough to care about my backache, listen to my frequent complaints, and provide impeccable consultancy on the interpretation of the COVID-19 pandemic in Chap. 11. Finally, the dedication of this book belongs to my daughter Anita and son Antonio and to my granddaughter Miranda, who are making life much richer. Of course, the responsibility for the product is only mine.

Bologna, Italy Francesco Taroni

Contents

1	**Health in the Making of a Nation**	1
	The Health of the Population	3
	The Piedmontization of the Health Legislation	8
	The Long Path to the Sanitary Code	10
	The Law of Public Health	12
	Secularizing Opere Pie	14
	Implementing Crispi's Laws	17
	The Giolittian Era: Reformism Without Reform?	19
	References	25
2	**Health and Healthcare in the Liberal State**	29
	Cholera, the Nineteenth-Century Disease	31
	The Rise and Demise of Pellagra: A Success Story	39
	Malaria and Quinine: The First Magic Bullet	43
	Doctors, Medical Care, and Work Safety	48
	The Rise of the Italian Hospital: Change Without Reform	54
	References	64
3	**The Great War and the Spanish Flu**	73
	A Public Health Disaster	75
	The Medicalization of War	77
	The Militarization of Psychiatry	78
	Disabled Soldiers and Social Reform	80
	Influenza Epidemic and the War	83
	References	86

4 **Health Under the Fascist State** 91
The Ascension Day Address: A Political and Ideological Agenda 92
Bonifica Integrale: Land Reclamation and Anti-malarial Policy 95
The Institutional Pillar: Carta del Lavoro and Casse Mutue 99
A World Apart: The Battle Against Tuberculosis 102
The Fragmented Fascist Healthcare State and Its Doctors 107
References 115

5 **Postwar: Roads Not Taken** 121
A Country in Ruins 122
Health and the War 124
A Lukewarm Transition 127
The Beveridge Report in Italy 130
The Constitutional Debate 133
The Rationalist Illusion: D'Aragona Commission 134
References 137

6 **The Rise and Fall of the Mutual Jungle** 141
A New Old Policy 142
The Ministry of "salvo" 147
The Dynamics of Spending and Service Provision 148
Hospitals: Insufficient, Inadequate, and Unprepared 151
A Legitimacy Crisis 158
The Reformist Mirage 160
References 162

7 **The Creation of the Servizio Sanitario Nazionale** 165
The Collapse of the Sickness Funds 166
The Regions, Laboratories of Democracy 167
The Season of the Movements 168
Passing the Law 170
Bugs and Holes of the Reform 173
A Reform Out of Time 177
The Undoing of the Reform 178
References 182

8 Reinventing the SSN? 187
The Origins of the "reform of the 1978 reform" 188
The Content: An Assemblage 190
A Swift Process 193
Implementation and Its Impact 195
"Rational rationing" of Drug Benefits 197
Back to Base: The 1999 Reform of the 1992 Counter-reform 198
A Turbulent Process 202
References 206

9 New Issues at the Dawn of the Twenty-first Century 211
The First Two Decades of the HIV/AIDS Epidemic 212
From the Politics of Access to the Policy of Excess 217
Responding to Public Expectations: Citizens' Views 224
References 226

10 A Lost Decade 231
Regionalization, Alleged and Real 232
Contractual Relations: Patti per la Salute and Piani di Rientro 240
References 244

11 Two Converging Crises 247
Weathering the Great Recession 248
The Second Storm: The COVID-19 Pandemic 257
References 269

12 Postscript 275
Regime Types and Health Policies and Institutions 276
Ruptures and Continuities 282
The Role of Organized Medicine 285
Population Health 286
Global Health and National Histories 287

Index 289

Abbreviations

ACIS	Alto Commissariato per Igiene e Sanità (High Commissariat for Hygiene and Healthcare)
AHCPR	US Agency for Health Care Policy and Research (now AHRQ, Agency for Health Care Research & Quality)
AMI	Associazione Medica Italiana (Italian Medical Association)
ANAAO	Associazione Nazionale Aiuti Assistenti Ospedalieri (National Association of Hospital Medical Assistants)
ANMC	Associazione Nazionale Medici Condotti (National Association Municipal Doctors)
CLN/CLN (AI)	Comitato Liberazione Nazionale (Alta Italia)/ Committee of National liberation (for Upper Italy)
CNEL	Consiglio Nazionale dell'Economia e del Lavoro (National Council for Economy and Labor)
CNR	Consiglio Nazionale delle Ricerche (National Research Council)
CPA	Consorzio Provinciale Anti-Tubercolare
ECB	European Central Bank
EMS	European Monetary System
ERP	European Recovery Plan
EU	European Union
EUR	Esposizione Universale di Roma (Universal Exhibition—EU42)
FNOM/FNOM CeO	Federazione Nazionale Ordini dei Medici (National Federation of Doctors' Orders)
HCFA	US Health Care Financial Administration (now CMS, Center for Medicare and Medicaid Services)

ICU	Intensive Care Unit
ILO	International Labor Organization
IMF	Internazional Monetary Fund
INAM/INFAM	Istituto Nazionale per l'Assicurazionei Malattia/Istituto Nazionale Fascista Assistenza Malattia (National (Fascist) Institute for Sickness Assistance/Insurance)
INAIL/INFAIL	Istituto Nazionale (Fascista) Assistenza Infortuni sul Lavoro (National Fascist Institute for Insurance in Industrial accidents)
INPS/INFPS	Istituto Nazionale (Fascista) della Previdenza Sociale (National (Fascist) Institute of Social Security)
IPAB	Istituti di Pubblica Assistenza e Beneficenza (Institutes of Public Assistance and Beneficence)
IRI	Istituto per la Ricostruzione Industriale (Institute for Industrial Reconstruction)
ISS	Istituto Superiore di Sanità (National Institute of Health)
ISTAT	Istituto Nazionale di Statistica (National Institute of Statistics)
LOH	League of Nations
MAIC	Ministero della Agricoltura, Industria e Commercio
NHS	(British) National Health Service
OECD	Organization for Economic Cooperation and Development
ONC	Opera Nazionale Combattenti (National Organization of Combatants)
ONMI	Organizzazione Nazionale Maternità e Infanzia (National Organization for Mothers and Children)
OTA	(U.S.) Office of Technology Assessment
SSN	(Italian) Servizio Sanitario Nazionale
UNRRA	United Nation Relief and Recovery Agency

List of Figures

Fig. 1.1	Crude death rate per 1000 population, 1872–1920. (*Sources:* Bizzozero G., 1900; ISTAT, 2011)	5
Fig. 2.1	Number of deaths from malaria, pellagra, and tuberculosis, 1887–1915, selected years	30
Fig. 2.2	Mortality from malaria, 1900–1914, Number of deaths per million population. (*Source*: ISTAT, 2011)	47
Fig. 4.1	Number of beds in sanatoria INFPS, 1929–1941 (in grey: new beds). (*Source*: Francioni, 1957)	105
Fig. 5.1	Postwar: crude mortality rates per 1000 population, 1939–1950. (*Sources*: ISTAT, 1949; Tizzano, 1950)	125
Fig. 6.1	Trend in pharmaceutical and hospital expenditures, INAM, 1960–1973. (Sources: Annali di Statistiche Sanitarie INAM, various years)	151
Fig. 6.2	Percentage of hospital admissions among persons insured with INAM, 1955–1974. (Source: INAM, Annuario Statistico, various years)	153
Fig. 11.1	Public health expenditure, per cent of GDP, Italy 2000–2018. (*Source*: WHO Global Health Expenditure database)	251
Fig. 11.2	Number of reported COVID-19 cases, February 2020–June 2021. (*Source*: Istituto Superiore di Sanità Open Data www.iss.it)	258
Fig. 11.3	Number of COVID-19 deaths, February 2020–June 2021. (*Source*: Istituto Superiore di Sanità Open Data www.iss.it)	259

List of Tables

Table 1.1 Number and expenditures of *Opere pie* in relation to their principal goal (broad categories), 1880 15
Table 1.2 "Social" expenditures by source, selected years (million lire) 21
Table 2.1 Number of deaths from the main waves of cholera epidemics in Italy, 1861–1911 31
Table 2.2 Pellagra: total cases, new cases, and deaths, 1879–1930, selected years 40
Table 2.3 Hospitals and their activity, 1885–1914, selected years 61
Table 4.1 Public and private hospitals by type, 1932 110
Table 6.1 Population covered by *casse mutue*, 1943–1974, selected years 144
Table 6.2 Distribution of public hospitals according to Petragnani categories, 1954 152
Table 11.1 Ten regions with the highest rates of reported COVID-19 cases as of June 1, 2021 259
Table 11.2 COVID-19 pandemic: tests, new cases, and deaths, first and second waves (average rates per week per 100,000 population) 260

CHAPTER 1

Health in the Making of a Nation

At its birth in 1861 the Kingdom of Italy was a poor and illiterate country of about twenty-two million people, a nation of regional diversities and strong municipal traditions, divided between city and countryside, and northern and southern parts. Forty per cent of the population lived in absolute poverty, and seventy-eight per cent were illiterate (Benini 1911). In 1861, the Casati Law and in 1877 the Coppino Law prescribed primary education free and compulsory for children between six and nine years, but responsibility for financing was left to dilapidated municipalities. Forty years after unification, education was still "the gloomiest chapter in Italian social history, a chapter of painful advance, of national indifference to a primary need" (King and Okey 1901, p. 233). Differences between cities and the countryside and the north and the south of the country increased. Illiteracy fell to 56 per cent, but was as low as 17 per cent in Piedmont and 21 per cent in Lombardy and remained at 70 per cent and higher in the south. It was only in 1911 that the Daneo-Credaro Law gradually transferred cost and responsibility from local government to the central state and made teachers state employees. Education became the most centralized public policy of the liberal state, although variation in resources and performance remained great.

Italy was largely an agricultural country. In 1861, about eight million Italians worked in agriculture and only three million in crafts and manufacturing. Agriculture displayed all the varieties of cultivation: intensive farming in the Po Valley plains of Lombardy and Emilia; sharecropping in

F. Taroni, *Health and Healthcare Policy in Italy since 1861*,
https://doi.org/10.1007/978-3-030-88731-5_1

Tuscany and central Italy; *latifondo* (large estate), producing cereals and raising cattle, in Lazio, Apulia, and the isles; *giardini* (gardening areas) for export products such as wine, oil, almonds, and citrus fruits in Sicily (Davis 2000, p. 235). Industry was weak, mostly textile, and accelerated its development at the turn of the century, particularly in the "industrial triangle" of Milan, Turin, and Genoa, and this increased the north-south divide (Zamagni 1993). The standard accounts of the Italian economy after unification typically described a trend of sluggish economic growth until the economic "spurt" of the Giolittian era, with a prolonged agricultural slump in the 1880s. Newer estimates describe a long-term growth rate uniformly higher over the entire period, where the end-of-the-century Giolittian spurt completely disappears and the supposed agricultural crisis of the 1880s turns instead into "years of prosperity" (Fenoaltea 2011, p. 6). Uncertainties about the pattern of economic growth in the first fifty years after unification do not help in understanding the association between economic growth and changes in population health.

Despite the predominance of agriculture and a delayed industrialization, a relatively high proportion of the Italian population lived in provincial towns and larger cities compared with the leading continental powers of Germany and France (Sassoon 2019, p. 46). The proportion of those living in the countryside decreased from 64 per cent in 1861 to 46 per cent in 1901, compared to 53 per cent in Germany and 62 per cent in France (Contento 1902). The population of *comuni* (municipalities) with more than 100,000 inhabitants doubled from 1.6 million to 3.1 million (Mortara 1908, p. 393).

Institutionally, Italy was a constitutional monarchy with a senate appointed by the King and a chamber elected by less than 2 per cent of the citizenry. In politics, the *questione romana* (the problem with the Catholic Church) took centre stage for a long time. The papacy refused to recognize the "Usurper State" and worked for its dissolution by acting as "a State within a State" (Pollard 2011; see also Pollard 2014). This had important consequences in the constant strive for "secularizing" *Opere pie*, the main agents of poor relief in the liberal age. Only later, towards the end of the century, the *questione sociale* (the social problem) and *la questione meridionale* (the southern problem) emerged on the political agenda.

The health of the population was poor. The "deluge of ancient and new evils that plague Italy" in its early years was vividly described by Senator Carlo Maggiorani. "Tuberculosis, scurvy, rickets hold the country more than ever; pellagra expands its ground; malaria's distressing effects sickens much of the peninsula [...]; syphilis meanders unruly [...]; deaths in children, soldiers and peasants outnumber our expectations, because of the weaknesses of the former and the behavior of the others; exotic infections are easy to catch and spread quickly; smallpox keeps raising its head; diphtheria unfolds and increases by day; epileptics and lunatics augment steadily; the abuse of liquor is ubiquitous" (Maggiorani Senate March 12, 1873; see also Della Peruta 1980). These were, in part, legacies of the seven pre-unification states. They were also the effects of the neglect of health policy by the liberal state. Health issues entered only belatedly in the national policy agenda, despite doctors' demands for political influence in the process of building the new state. It took a long and convoluted path to pass Crispi's sanitary code, and its implementation was patchy. This is the main subject of this chapter, after a quick look at the health of the population, which the next chapter examines in more detail.

The Health of the Population

The dearth of good statistical information concerning the health of the Italian population is but one of the aspects of the neglect of health issues in the policies of the liberal state. Cause-specific mortality data are only available for the whole country beginning in 1887 and were recorded starting in 1881 only for the 281 major *comuni* out of over 8000, covering about 7 of the over 28 million inhabitants (MAIC 1880, p. 187). The inadequacies of health statistics contrasts with the interest in statistics as an instrument of nation building and a tool for a "national pedagogy" to make the Italians that had begun with the *Congressi degli Scienziati* in the Risorgimento (Patriarca 1996; Favero 2001). However, interest remained high: private initiatives complemented official surveys, and professional institutions funded prizes for dissertations, which, at times, became part of official statistical series. This was the case, for example, of Giuseppe Sormani's *Geografia nosologica dell' Italia* (Nosological geography of Italy) winner of the 1887 prize of the *Regio Istituto Lombardo di Scienze e Lettere* of Milan, which was eventually published by *Annali di Statistica*, and is now the principal source of cause-specific mortality data before 1881 (Sormani 1881).

Significant information about the health of the population also comes from literary sources, as well as from the outburst of official enquiries of the early 1880s, which explored a host of issues, from education to agriculture, industry, municipal services, and so on, with an eye to standard of living and health-related behaviours. Asking for an in-depth national survey, Agostino

Bertani denounced the harsh status of the peasants who were "yellow, emaciated, dirty, and downcast by misery, malaria, fevers and social neglect" (Bertani Chamber June 7, 1872). In Veneto, part of Lombardy, and Emilia-Romagna, the rural population was afflicted by pellagra, because of their diet of *polenta* (maize bread). Malaria was endemic in the rice fields of Lombardy and Emilia, the coastal swamps and flooded areas of central and southern Italy, such as Pontine Marshes around Rome, but also in Campania, Apulia, and Calabria, as well as the isles of Sicily and, particularly, Sardinia. Stefano Jacini's *Inchiesta Agraria e sulle Condizioni della Classe Agricola* (Inquiry into Agriculture and the Conditions of the Agrarian Class) documented their poor standard of life (Jacini 1885). In most parts of Italy peasants lived in overcrowded dwellings, which they shared with their domestic animals, and consumed a miserable diet, poor in nutrients and vitamins, consisting of bread made with maize or chestnuts and a few vegetables. Pasta, wine, and a little meat were consumed only rarely and by the more prosperous (Celli 1896; see also Sassoon 2019, p. 55). The urban population did not fare much better. Matilde Serao (1884) and Jessie White Mario (1895) described the dire conditions of the Neapolitans living in the *fondaci* of the *città bassa* (lower city). In his *Lettere Meridionali* Pasquale Villari, senator and several time minister, compared them with life in outcast London, and found it much worst (Villari 1878). Municipal "poor registers", that is, the list of people eligible for relief by municipal authorities, usually covered about one-third of the city population, and grew to 50 or even 60 per cent in times of crises, such as increases in food prices, poor harvests, or epidemics (Woolf 2000, p. 431).

Available data suggest that mortality remained stable for about twenty years after unification, at approximately the rate observed in most pre-unification states of about 30 deaths per 1000 population (Sori 1984). It is hard to determine exactly when mortality began its decline and why. Giulio Bizzozero provides crude mortality rates per 1000 population from 1872 to 1887, which are slightly over 30 deaths per 1000 from 1872 to 1875 and stabilize around 27 deaths per 1000 in the 1980s (Bizzozero 1900, p. 607). The 1890–1900 decade is the first period for which reliable data are available for the whole country. In the 1990s, mortality started to decline steadily up until the First World War (Fig. 1.1) (ISTAT 2011). However, by the end of the century, Italy still lagged leading European countries. In the decades 1884–1893, the crude mortality rate of 26.8 per 1000 population observed in Italy compared poorly with 22.4 in France, 24.5 in Germany, and 19.1 in England (Celli 1896).

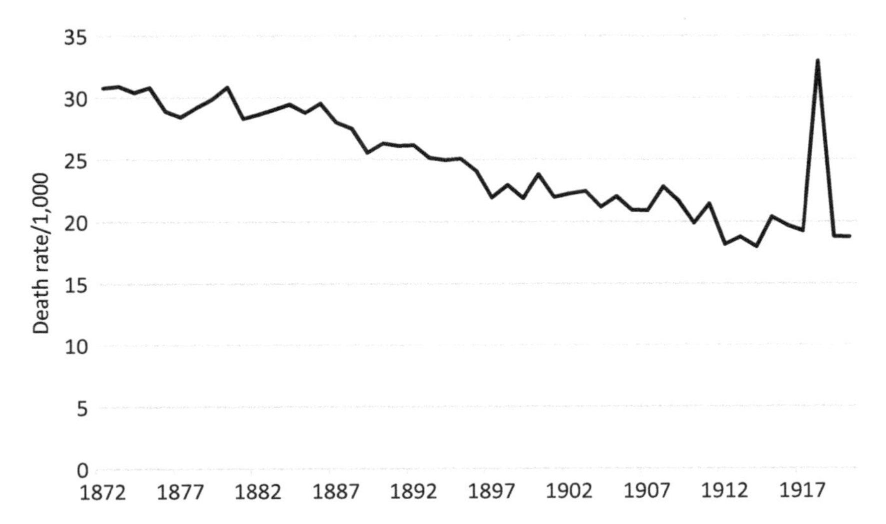

Fig. 1.1 Crude death rate per 1000 population, 1872–1920. (*Sources:* Bizzozero G., 1900; ISTAT, 2011)

The improvement in mortality started earlier, fell faster for infectious diseases, and was concentrated particularly in children. Infectious disease as a group accounted for about one-third of total deaths in the 1880s, but by 1900 their contribution to mortality was lower than 20 per cent. Mortality rates declined more rapidly in the cities than in small towns and rural areas. The Italian situation contrasted with that in Victorian England, where the urban environment was generally more unhealthy than the countryside, the so-called urban penalty (Woods 2003). Anthony Wohl pointed out that "it was not that rural conditions were good but that many elements necessary for good public health were much more difficult to provide in the congested and burgeoning cities" (Wohl 1983, p. 3). Several explanations of this inversion were provided. For example, the *Annuario Statistico* of 1886 reported that typhoid deaths were more frequent in small *comuni* than in big cities "which had benefited from the good waters and the improved housing provided in the last twenty years" (MAIC 1886a). This positive appraisal contrasted with the results of the *Indagine sulle condizioni igieniche e sanitarie nei Comuni del Regno* (Enquiry into the hygienic and health conditions of the communes of the Kingdom) published in 1886, in the midst of a cholera epidemic. The survey documented that in the early 1880s over fourteen million people, more than half of the Italian population, lived in places lacking any form of sewage disposal, while only about one-third of the communes had some type of underground water mains (MAIC 1886b; Frascani 1980). In any case, the anonymous author of the official government publication seemed to find solace in the idea that preventing the current "148,000 deaths for similar infectious and miasmatic diseases (18.8 percent of total deaths)" would greatly benefit the country (MAIC 1886b, p. 481). Years later, commenting on the decline of deaths for typhoid and diphtheria in the big cities, an anonymous reviewer of the *Giornale della Società Italiana d'Igiene* also stressed "the significant improvements in housing and drinking water which had been achieved in the last few years" (Anonymous 1892, p. 81). He also noted, however, that mortality for gastroenteritis and diarrhoea in Italy was still "enormous" and remained stable at rates over 30 deaths per 10,000 population, well over France (21), Prussia (10), and England (7), concluding that those waterborne conditions "mostly contributed to keeping the Italian overall death rate higher than that observed in many other European countries" (p. 85). The big cities of Rome, Milan, and Turin showed a lower mortality than their surrounding countryside up the turn of the century (Mortara 1908, p. 403). Their mortality however remained higher than other big European cities such as Berlin, Paris, and London, which Mortara attributed to the Italian cities being more ancient, which made sanitary works more difficult (p. 405).

Mortality declined in all age groups but children under five showed the greatest reduction, which explains most of the increase in life expectancy at birth observed between 1881 and 1901, from 35.5 to 42.6 years (Somogy 1967). Infant and child mortality rates were lower in the northern regions of Piedmont and Veneto than in the south, particularly Apulia, Basilicata, and Calabria. Deaths from typhoid fell, but bronchitis, gastroenteritis, and diarrhoea remained the leading causes of death in children under fourteen years. Tuberculosis was the largest single cause of death, leading to about 60,000 deaths per year between 1887 and 1914. Its rate however declined from 21.3 to 17.0 per 10,000 population in 1905–1907 (Benini 1911), much lower than those observed in France. The pulmonary form of tuberculosis was most prevalent in the big cities of the northern regions, while in the south abdominal forms were more frequent, along with deaths from simple pneumonia. Pulmonary tuberculosis was typically associated with the industrializing city with its overcrowded housing and occupational exposures. The predominance of abdominal forms in the south may suggest poor surveillance of contaminated milk and/or meat, possibly associated with underreporting of cases of pulmonary tuberculosis because of lack of clinical skills, availability of diagnostic facilities, or just poorer reporting. In the late 1880s, deaths from malaria and pellagra, Italy's "twin scourges", were about 20,000 and 5000 respectively, and their numbers were also declining, as we will see in the next chapter. In summary, the first phase of the health transition in the last two decades of the nineteenth century was slow and characterized by a sharp north-south gradient. The gradient was strongly conditioned by infant and child mortality rates, mostly due to infectious diseases such as measles, diphtheria, and gastroenteritis. Mortality for tuberculosis was also crucially important. Deaths for pulmonary tuberculosis were largely prevalent in the northern cities and were lowest in Basilicata, Calabria, and Sicily.

At the turn of the century, Italy had not yet fully exploited the opportunities that increased medical knowledge and the bacteriological revolution offered to improve the health status of its population. Its belated epidemiological transition put Italy well behind other major European countries in terms of mortality levels and disease patterns. Along with the other key indicators of "civilization", such as violent crime and illiteracy rates, the "sad statistics" of mortality (Pareto 1891) revealed the backwardness of Italy and its poor performance compared to the leading nations and served as a call to reform its health institutions (see, e.g., Celli 1896). In addition, the mortality gap between the big cities and the countryside and between the northern and southern parts of the country raised the question of the autonomy of the municipalities and of their relations with the central government. National elites took the poor conditions of the rural communes, particularly in the south, as a sign of the inadequacies of local authorities. Patriarca reports that local state officials often complained of municipalities' neglect of their assigned duties of diffusing elementary education, building roads, and improving sanitation (Patriarca 1994, p. 228). This contradicted the implementation of the liberal idea of "self-government", which prompted liberal governments of the right and of the left to assign communes an increasing number of unfunded "modernizing" projects. However, communes could successfully perform only with adequate finances, human resources, and professional skills. The significant decline in waterborne diseases can be explained primarily by the surge of sanitary works undertaken by major cities in the wake of the cholera epidemic of 1884–1887, when the state dispensed to *comuni* massive long-term loans to finance capital projects for slum clearance, public housing, and the building of water mains and sewers (Giovannini 1996). Moreover, the Giolittian era's transformations of municipal institutions governing the public services of modernizing cities, such as electricity, gas, water, and transports, were only partially associated with improvements of their public health departments under the direction of the new post of *ufficiale sanitario*. This also explains how the complex mix of socio-economic events and sanitary and medical interventions occurred at different times in various places and spread at variable speed across a country that was still characterized by regional diversities and strong municipal traditions. This was, in part, the legacy of the seven pre-unification states and the effects of the wide differences among the various parts of the country. It was also a problem of the health policy of the liberal state, including the choice of "piedmontization" as a means of nation building. The decision to impose on the rest of the country the backward health legislation of Piedmont frustrated the expectations of doctors and their demand for political influence and was a material setback for physicians who worked in areas of the country such as Lombardy and Tuscany, which enjoyed more advanced health legislation.

The Piedmontization of the Health Legislation

Doctors were the professionals that made one of the greatest contributions to the Risorgimento: they manned the barricades in 1848, they fought the wars of independence, and many of them were exiled for their patriotism (Malatesta 2011, p. 65). Participating in the patriotic and scientific debates of the *Congressi degli Scienziati* (Congresses of Italian scientists), doctors had represented themselves as the "hygienic conscience of the Nation" and the "true" experts concerning the "real" health needs of the population, recognized by the progress of scientific medicine (Pancaldi 1983). However, at unification their great expectations were disappointed by the legal transplant of the backward Piedmontese Rattazzi law of 1859 to the annexed territories of the Kingdom. The Italian Parliament received a deluge of *petizioni* (claims) demanding participation in the process of nation—building from the annexed provinces and professional organizations, including medical doctors. Doctors were particularly incensed in Lombardy, where the transplanted Piedmontese legislation jeopardized relief for the poor and public health and endangered the economic position and social standing of *medici condotti*, the municipal doctors who had traditionally provided free care for the poor.

While in Lombardy *condotte municipali* had ensured medical care for the poor since 1834, the Piedmontese Rattazzi law of 1859 did not recognize *condotte mediche* as obligatory expenses of *comuni*, and worse still, the new Governor of Lombardy, Senator Vigliani, revoked the *Statuto austriaco* (Austrian charter), which granted medici condotti hard-won occupational stability. Gaetano Strambio was a liberal progressive from Milan and a long-time leader of Italian doctors. In the authoritative journal *Il Politecnico* he called the Rattazzi law "a permanent attack against the health of the public and a hateful way of dealing with doctors" and called the Governor's decree " the product of the most arcadian ignorance" (Strambio 1862, p. 255 and 248). Doctors' activism through *Associazione dei Medici Italiani* (AMI—Italian Medical Association) and professional journals however did not accomplish much (see for example Castiglioni 1866).

The administration of the new Italian State was consolidated in 1865 with the *Legge per l'Unificazione Amministrativa del Regno d'Italia* (Law for the Administrative Unification of the Italian Kingdom, n.2248 of 20 March 1865). The law reinstated as compulsory all municipal expenditures (except for drugs) for "medical, surgical and midwifery" services for the poor. However, its Annex C put public health under the political authority of the Minister of the Interior, through its Prefects, Under-Prefects, and Mayors, providing only a consultative role for doctors. The legislation defined a residual system of poor relief where *beneficenza pubblica* provided residential care through charitable, mostly religious, organizations, while domiciliary support of *medici condotti* and midwives was ensured through *carità legale* (legal charity) legislated by the state but administered and funded by *comuni*, acting as providers of last resort for their resident poor. *Carità legale* applied only to the citizens of the commune and was left to the allocation of municipal budgets. They were not based on purpose taxes, such as the poor rates in English parishes, but were financed by excise taxes on consumption goods, such as the salt tax and the much-hated *tassa sul macinato* (grist or bread tax), which disproportionately hit the poor. The law introduced the fundamental financial principle of the liberal state that public spending for local services, including education and health, was legislated by the central state but left to municipalities to fund. The decentralized system of financing public expenditures established a human capital trap for the poorest areas of the country with higher needs, fewer resources, and less capability that contributed to the huge disparities between the cities and the countryside and the northern and southern parts of the country. *Società di Mutuo Soccorso*

(Societies for Mutual Aid), similar in principle to the Friendly Societies in England, were of limited impact in health insurance, as their key benefit was income replacement and they were few in number, of small size, and mostly concentrated in the largest and industrial cities of Milan and Turin (Marucco 1980).

Governments of the *Destra Storica* (Historic Right) are credited with balancing the national budget, which was burdened by a substantial public debt of about 40 per cent of the gross domestic product (Zamagni 1993). The public debt was consolidated into the *Gran Libro del Debito Pubblico* (Great Book of the Public Debt) and attacked with a strict policy of fiscal austerity, or of "economy to the bone", mostly associated with Finance Minister Quintino Sella, based on high taxation and low expenditures, with obvious implications for social programmes. Between 1862 and 1880, tax revenue more than doubled and Italians became the most heavily taxed people in Europe. State expenses for health and social issues remained well below 1 per cent of the Ministry of the Interior's budget, or around one million lire, while provincial and communal expenses amounted to about seventy million lire (Battilani 2000).

The Long Path to the Sanitary Code

The first step in the twenty-year-long quest for comprehensive public health reform was Prime Minister Bettino Ricasoli's promise of a new *Codice Sanitario* (Sanitary or Health Code). The terms of reference of the Commission appointed in 1866 warned the members "to protect personal freedom to the maximum extent and avoid damaging, or just jeopardizing, free usage of property and personal resources" (cited in Ognibeni 1982, p. 602). A new Commission appointed seven years later was similarly instructed by Prime Minister Giovanni Lanza to "bear in mind public economy, industry, trade, agriculture, and public order. All these factors have the right… of not being constrained for mere hygienic reasons and in absence of absolute necessity". The problem of harmonizing hygienic practice with personal freedom and private property emerged most clearly in debating the measures to prevent overcrowding in urban housing. The *relatore* (speaker) for the law Senator Carlo Burci rhetorically asked "but how can we enter private places? How can we order that rooms with a capacity of four accommodate just four people? And in addiction, what if the owner would like to admit eight people: should we go and inspect the place? Moreover, how could we perform the inspection? And from where do we get the right to do that?" (cited in Ognibeni 1982, p. 602). Disease surveillance, notification of infectious disease, and smallpox vaccination raised similar problems of privacy and freedom, not unlike the issues which were also debated in England (Hanley 2016). The bill was buried with the end of the legislature, but the debate had made clear that the allegiance to private property and personal freedom were the main barriers in the way of a Sanitary Canon. A policy based on different principles required a change in politics that materialized only when the *Sinistra storica* came to power and Prime Minister Agostino Depretis asked his old friend Agostino Bertani to draft a *Codice della Pubblica Igiene* (Public Health Code). This was the turning point in the long path towards new public health legislation.

Agostino Bertani was a patriot, soldier, and physician who had served Garibaldi with Depretis and was the political leader of *Radicali*, a parliamentary group of the *Estrema* (Extreme left). He took almost three years to draft his Code, which was by any measure radical in principle and innovative in organization (White Mario 1890). The dominant principle of the Code was that "public hygiene ought to be *ordered*" (p. 355, italics in the original). His sources of inspiration were France and Germany, but particularly England because there "everything always follows the sanitary law and *salus publica* is really *suprema lex*" (italics in the original). The basic unit of the new organization was the *Servizio Igienico-sanitario Comunale* (Municipal Hygienic and Sanitary Service), a multiprofessional team of *medici condotti*, midwives and veterinarians. At the national level, the *Magistrato Superiore della Pubblica Igiene* (Higher officer for public hygiene) "sat next to the Minister of the Interior" and was supported by *Consiglio Superiore di Sanità* (Higher Board of Health), whose members were elected by doctors' organizations. Medical doctors lead each level of the organization since "the surge of hygiene to science" would meet the health needs of the population. Bertani's Code of Public Hygiene never became law, but his radical propositions left a lasting imprint on the debate. The aphorisms *la sanità pubblica deve essere comandata* (public health must be ordered) and *salus publica suprema lex* (public health is the highest law) anticipated a shift in the language of public health away from the ideology of laissez-faire liberalism and became the buzzwords legitimating state intervention in public health for the years to come.

The Law of Public Health

Francesco Crispi's "parliamentary dictatorship" (as defined by Denis Mack Smith) during his first Cabinet from 1887 to 1891 hastened the enactment of sweeping legislation aimed at "reforming the State to build the Nation" (Duggan 2002). The ravaging cholera epidemic of 1884–1887, which exposed the dismal hygienic conditions of the country and its poor management, offered a window of opportunity to force a radical and comprehensive reform in public health. The public health law largely followed Bertani's principles, which Guido Bizzozzero, the saviour of the Turin School, widely shared with Luigi Pagliani, the professor of hygiene of the same University who masterminded the Code and its associated legislation (Pagliani 1901). The institution of an autonomous *Direzione di Sanità* (Health Directorate) within the Ministry of the Interior transferred the governance over sanitary matters from political and administrative to a medical body. Its institution by Royal Decree in 1887 was part of the general administrative reform of the Ministries, and the new Directorate took responsibility for the disparate features of health policy previously dispersed across various Ministries. Intellectually, the *Direzione di sanità* meant continuity with Bertani's *Magistrato superiore di sanità* sitting next to the minister. In practice, *Direzione di sanità* with its two chemical and microbiological Laboratories and the School of Public Health for training the new officers of health envisaged by the impending legislation was a vital facility for a modern organization based on science and led by medical doctors (Pagliani 1889). In a detailed Report to the *Consiglio Superiore di Sanità*, Pagliani duly documented universities' shortfalls in teaching and researching the rapidly evolving empirical science of hygiene (Pagliani 1890). This led to the worsening of his relations with the University of Rome, which resented the expropriation of the Laboratories and the concentration of power in Pagliani's hands. Despite mounting opposition, the final step on the long path towards the health code went smoothly, and the Law of December 22, 1888, nr. 5848 "*Tutela della Igiene e della Salute Pubblica*" (Protection of Hygiene and Public Health) was passed in both houses in record time.

The law established a three-layered system, centralized under the Ministry of the Interior, backed by *Consiglio Superiore di Sanità* (National Health Board) and the technical governance of the *Direzione di Sanità*. *Condotte mediche* were the backbone of the new system, a multiprofessional team composed of *medici condotti*, midwives, and, where required, veterinarians, providing domiciliary medical and obstetric care for the poor and control of zoonotic diseases in agricultural districts. *Ufficiali sanitari* (municipal officers of health) where available, or *medici condotti*, organized surveillance of infectious disease, supervised sanitary works, controlled the location and pollution of factories, organized inspections of housing, and foodstuff, particularly meat and milk. To those objecting the idea that *medici condotti* could take the function of *ufficiali sanitari*, Mario Panizza (the speaker of the law and formerly Bertani's close collaborator) refuted their objections by noting the small number of qualified hygienists available and the burden on municipal coffers, also arguing for the benefit of combining preventative and curative medicine. At the intermediate, provincial level stood *Medico provinciale* (provincial medical officer), appointed by the Prefect and acting as a link between the centre and the municipal level. The law took pride in grounding the organization of public health in the new science of bacteriology and set up a network of public health laboratories and vaccination centres in municipalities with a population of over 20,000, backed by national chemical and microbiological laboratories to process the samples sent by *medici provinciali* and *ufficiali sanitari*. Panizza remarked that the key innovation of the law was that "all executive branches of the health administration were entrusted to technical personnel" (Panizza Chamber December 13, 1888). In fact, the law left many of the doctors' traditional demand for professional autonomy and occupational stability unaddressed. *Medici provinciali* and *ufficiali sanitari* maintained their advisory role to the mayor and the Prefect. Medical members of the Health Boards were still nominated instead of being elected by their peers. The three-year appointment period of *medici condotti* continued to expose them to mayors' abuse of their power of removal.

The principal issues with the law were the huge administrative tasks and financial burden placed on *Comuni*. The financial implications of the law had had a very small part in the parliamentary debate. Additional expenditures were roughly estimated at about forty million lire, of which thirty-nine million lire would come from municipal budgets (Imperatori 1891b, p. 379). This would further increase the imbalance between central and local expenditure and put additional stress on municipal fiscal capacity. *Comuni* rarely had the manpower and technical expertise to implement the push to modernize the practice of hygiene and home healthcare, except where more advanced organizations were already in place, such as in parts of the Lombardy, Veneto, and Tuscany, and this could again increase inequalities across the country. The School had taken on itself the mammoth task of training the new medical and technical bureaucracy, admitting to its six-month courses for medical doctors, as well as engineers, chemists, physicists, and veterinarians (Pagliani 1889), but its resources were clearly less than adequate.

Crispi and Pagliani's public health law was well received in Italy and abroad. The *British Medical Journal* celebrated the Italian "resurgence of public health" (Anonymous 1887, 1889). In Italy, it was praised from the right and from the left of the political spectrum as an exemplar of the renewed grandeur of Italian medicine, which put Italy among European great powers. Many, including Benedetto Croce, even associated the law with "the disappearance or diminution of epidemics and other diseases and the decline of deaths" (Croce 1929, p. 180). Chief criticisms came from a series of papers published by the liberal (at the time) *Giornale degli Economisti*, which was highly critical of Crispi's policy and of the law's principles and contents. Ugo Imperatori described Crispi's law as "one of the least brilliant illusions of State socialism" (Imperatori 1891a, p. 247) and concluded his long and detailed review wishing that "the actual order of the capital and under current political principles, reforms so seriously wrong should not be allowed and must be blocked" (Imperatori 1891c). Crispi's public health law, along with the law on *Opere pie*, has been a cornerstone of the Italian health organization for nearly a century. The medical profession rose to a position of cultural authority and gained influence in the process of policymaking, contributing to the modernization of the state and its administration, as demonstrated by the debate over the management of the cholera epidemics of 1884–1887, with which the law was in many ways intertwined (see page 31).

Secularizing Opere Pie

Opere pie (literally, Pious works) were a vast and fragmented assemblage of mostly religious charities of medieval origins created for the relief of the poor and the moral uplifting of beneficiaries as well as the donors. The first 1861 Survey listed 16,257 autonomous organizations. They were highly heterogeneous in aims (from curing the sick to providing dowries and paying ransoms), beneficiaries (the elderly, the chronically ills, the incurables, orphans, foundlings, single mothers, sick, and disabled people), and forms of relief (cash, in kind, such as food and clothing, and services, provided either indoor or outdoor). They also differed in size, financial resources,

Table 1.1 Number and expenditures of *Opere pie* in relation to their principal goal (broad categories), 1880

	Opere pie		*Expenditures*	
	Number	*%*	*Million lire*	*%*
Worship & Beneficence	6432	29.4	8.7	9.8
Charities	3800	17.4	2.8	3.1
Dowries	3035	13.9	1.4	1.6
Home assistance	2165	9.9	2.2	2.5
Hospitals, incl. asylums	1260	5.8	34.9	39.2
Foundlings & Children	1017	4.7	25.6	28.7
Education & Schools	801	3.7	1.7	1.9
Others	3356	15.3	11.8	13.2
	21,866	100.0	89.1	100.0

Source: Silei 2003, p. 111 and 114

and geographic distribution. *Opere pie ospitaliere* (hospitals) and other health-related organizations accounted for about 10 per cent of the number but more than one-third of total gross expenditures. Their number grew to 17,870 in the 1870s and further increased to 21,866 in the 1880s (Villari 1890), with a capital fund of about 1897 million lire, 40 per cent higher than that at unification (Table 1.1). Charities were concentrated in a few regions (Lombardy, Liguria, and Piedmont had about half of the capital) and in the largest cities, while half of the small comuni had no charities whatsoever. Total annual income (including returns on capital, occasional donations, and municipal transfers) amounted to about 135 million lire, although about half of the *Opere pie* declared annual income below 5000 lire. In the 1880s, their total annual net expenditures were approximately eighty-nine million lire, matching municipal social expenditure at about 0.7 per cent of the Gross National Product (Battilani 2000).

The vast, heterogeneous, and fragmented assemblage of the charities was described as "an immense and frightening forest" that resisted even the most strenuous efforts at penetration and control (Gabelli 1890). Reference was frequently made to their inefficient management and susceptibility to patronage and corruption. In his report from Sicily, Leopoldo Franchetti wrote, for example, that "for the honest ones they are a means of influence and favoritism, for the less honest a source of easy profits and illicit earnings" (Franchetti 1877). This explains the number of attempted reforms and why they inevitably failed. Francesco Saverio Nitti used the usual metaphor of the forest ("a shadowy forest that no one dares to enter") and explained that "they would not be laicized or modernized easily because of their medioeval origins, fierce independence, chaotic organization, ancient traditions, cherished privileges" (Nitti 1892).

The first reform was the so-called *Grande Legge* (Great Law) of 1862, essentially a "Nation's unification law" to provide a common legislative framework, and also an attempt to foster administrative and managerial efficiency. According to its *Relazione*, the law adopted the two principles of freedom and devolution, both supporting their self-government (Silei 2003, p. 29). Freedom was interpreted as following donors' wishes inscribed in charities' *tavole di fondazione* (a sort of constitution of the charity) because "the government has not the right to regulate the private use of personal wealth". Devolution aimed at preventing the intrusion of the central government in local affairs, robbing charities and the municipalities of resources for "their" poor, as the centralized, and redistributive, Bourbon legislation allegedly did. The external auditing of charities' balance sheets (but not of their budgets) was expected to foster the efficient use of their resources reducing the burden on the public purse in time of strict austerity. Even in relation to its modest aims, the law was widely perceived as a failure: an 1877 report found that 40 per cent of the balance sheets were still due. What ignited the debate and prompted a new quest for reform were estimates that *Opere pie* spent only about one-third of their annual gross income on charity, while the large majority allegedly were religious expenses (Villari 1890, p. 7). The appointment of yet another Commission, the Royal Commission of Enquiry chaired by Cesare Correnti, precipitated a harsh confrontation between the state and the Church. The fury of *Civiltà Cattolica*, the journal of the Jesuits and a semi-official voice of the Papacy, was aroused by the Commission's terms of reference, which included the task "to study and propose a plan for a general revision reflecting the spirit of the time and the new social conditions". The point that Civiltà Cattolica eloquently made against "the evil intentions" of "the usurper state" was that "the spirit of the time" had nothing to do with charities' "super-natural goal, dictated by Catholicism" for which they were answerable only to the Holy See and not to the secular State (Anonymous 1880, p. 12). Nine years later, however, the Royal Commission reached the unexpected conclusion that general expenses were only 12 percent of their gross income, which had greatly increased, as their capital: the Great Law of 1862 was "better than its fame" and than most of the laws one could find in other countries and only minor adjustments here and there were needed. Giuseppe Scotti, a member of the Commission and editor of the influential *Rivista di Beneficenza Pubblica*, the semi-official journal of the administrators of the *Opere pie*, triumphantly concluded that a reform was "neither prudent nor useful" (Scotti 1889, p. 5).

The minimalist recommendations of the commission did not satisfy Crispi. He reiterated his goals of "wrenching from the hands of unfaithful and covetous administrators the patrimony of the poor" "to nationalize charities and … incorporate them into the property of the State" (Crispi Chamber 30 Nov. 1889). The law of July 17, 1890, nr.6972 *Sulle Istituzioni Pubbliche di Beneficenza* secularized *Opere pie* imposing on them the legal status of *Istituzioni di Pubblica Beneficenza* (Institutions of public beneficence); amalgamated those with similar aims (a procedure named "grouping"); imposed new objectives when the original goals were no longer appropriate ("transformation"), while charities with a yearly income of less than 5000 liras were taken over by municipal *Congregazione di carità* ("suppression"). All charities were put under the governance of *Congregazioni di carità*, whose members were appointed by municipal councils (excluding the clergy and unmarried women). The law did not specifically address hospitals, except for imposing a duty to provide the facilities local universities needed for their medical teaching, but as we will see below it nevertheless spurred their medicalization. The expanded powers of the *Congregazioni di carità* made the governance of charities an important issue of local politics, which was particularly interested in hospitals as a matter of civic pride, source for employment and business, and opportunity for patronage.

Implementing Crispi's Laws

The implementation of Crispi's ambitious programme was blocked by the *crisi di fine secolo* (crisis of the end of the century), a defining moment in the political and institutional history of the country (Levra 1982). A chaotic combination of economic, financial, and political events transpired in a relatively short and dramatic period. Failures and scandals in the banking system, which implicated both Francesco Crispi and Giovanni Giolitti and touched the King himself combined with re-emerging problems in balancing the public budget. In 1896, the humiliating defeat in Adua halted the colonial adventure in Africa and brought down Crispi's second government. An agricultural crisis contributed to widespread popular unrest, starting in Sicily in 1893 and culminating in 1898 with the killing of protesters in the bloody *disordini del pane* (bread riots) in Milan, Turin, and Florence, followed by the imposition of martial law, the military government of general Luigi Pelloux, and the assassination of King Umberto I to avenge the victims of the 1898 repression. Some contemporary political observers worried about the future of the Nation, asking an anxious *Dove andiamo?* (Where do we go?) (Villari 1893), and others responded with *Torniamo allo Statuto!* (Let us go back to the Statute!) implying the abolition of the parliamentary regime and reinstating the full power of the Crown (Sonnino 1897).

The chaotic situation halted the implementation of Crispi's broad and comprehensive reforms. Barriers came also from multiple quarters. The Church rose against Crispi's attempt at secularizing charities with Pope Leo XIII's encyclical De Rerum Novarum, the charter of social Catholicism as its subtitle "On the condition of the working class" suggests, combating both liberalism and the rising socialism. This stimulated the grass-root movement of the *Opera dei Congressi* and the participation of Catholics in municipal elections, now of great political import for the administration of local charities (Pollard 2014, pp. 41–42). Locally, the tardiness of *Congregazioni di carità* in preparing their plans for concentration or outright suppression of designated Opere pie was consistent with local notables' reluctance to relax their hold on management. In addition, the cumbersome legal process of transformation required by law overwhelmed the occasional plan into a sea of legal skirmishes (Quine 2002, p. 58). The case for the implementation of the public health reform was a different story. The Crispi-Pagliani reform of public health envisioned a comprehensive system with the dual goals of reorganizing the provision of sanitary and domiciliary services and of building a research and educational infrastructure for the new system. However, both the Laboratories and the School were vehemently attacked by the Academy, and particularly by the medical faculty of the University of Rome. Its members were well represented in the Parliament and held important positions in the health administration. The exemplar character is Guido Baccelli, professor of clinical medicine at the University of Rome, deputy, three times Minister of Education and Chairman for life of the *Consiglio superiore di sanità* (Borghi 2015). In a long and heated debate in both Chambers the most prominent leaders of the Rome University, including Baccelli, Stanislao Cannizzaro, and Corrado Tommaso-Crudeli, did not speak against Crispi's law but strongly argued against the School and the Laboratories. The main arguments were that the School was infringing on the preserve of the University in education and the Laboratories had entered a field, such as the production of vaccines, better left to free market entrepreneurs to help the industry flourish. As a result, the School and the Laboratories were returned to the University of Rome, and the education of *medici provinciali, ufficiali sanitari* and their technical staff left to any willing University. Moreover, the *Direzione di sanità* was later downgraded to a simple *Ufficio* (Office) under the Direction of civil affairs in 1896 by Di Rudini's conservative cabinet. The technocratic drive of Giovanni Giolitti's cabinets resurrected the *Direzione di Sanità* some years later, and the Laboratories became the core facilities of the *Istituto di sanità* established in 1929, while the School was never reinstated.

Despite the major problems with their implementation, Crispi's laws established the structural and ideological foundations of health and social systems for much of twentieth-century Italy. The two laws were passed in the midst of the bacteriological revolution and in the aftermath of the debates over the "modern" management of the devastating cholera epidemic of 1884–1887, which opposed Luigi Pagliani and the Turin School to most of the academic establishment (see page 32). Both laws changed the organization and the practice of medicine in Italy and established the cornerstones of the two parallel systems of public health and domiciliary and hospital healthcare. Marginally amended during the Giolittian period, this asset was perfected by the fascist regime and retained for most of the republican period, up to the institution of the *Servizio sanitario nazionale* in 1978. The law on the *Opere pie* was not aimed at reorganizing hospitals, which was postponed and left to further legislation that was never passed. However, the secularization of the *Opere pie* under *Congregazioni di carità* paved the way for modernizing and medicalizing *Opere pie ospitaliere (see page 60)*. This exposed the main shortcoming of both laws. As Charles Rosenberg famously remarked for American hospitals, while the hospital was "medicalized" the medical profession became "hospitalized" (Rosenberg 1987). Since both laws simply ignored hospital medicine and its relations with domiciliary care, the rising of the hospital inexorably split hospital practice from home and community care.

The Giolittian Era: Reformism Without Reform?

The period from the end of the century to the First World War was one of industrial development and economic growth, as well as political and institutional change. At the turn of the century, the rapid growth of industry, particularly in the "industrial triangle" of Milan, Turin, and Genoa, increased production in the strategic sectors of iron and steel, electrical engineering, and automotive, and drove the growth of the Italian economy at a sustained rate of at least 5 per cent per year. Agriculture, still the dominant sector of the economy, also experienced a boom, because of its mechanization and state-subsidized programmes of land reclamation. Moreover, the balance of payment was boosted by the remittances from the millions of emigrants, particularly in the Americas (Toniolo 2013, pp. 37–68). The dominant political figure of this period was Giovanni Giolitti. Over the course of his political career, he served as a deputy uninterruptedly from 1882 to 1928. He was five-time head of government (a post he held along with the vital portfolio of the Minister of the Interior) from 1901 to 1914, with only short breaks, and also in 1892–1893 and again in 1920–1921, just before Benito Mussolini seized power. The so-called Giolittian era is generally seen as a period of social progress and modernization. Giolitti's governments extended the franchise to (nearly) all males, acknowledged workers' political and social rights, and maintained neutrality in labour disputes. They also modernized the administration, introducing new tools such as "special laws" and "special offices" supporting the aims of the interventionist state and providing "expert management" of social problems, such as emigration and labour. The institution of the Inspectorate of Labour and of the Council on Emigration were first "experiments" in "parallel administrations" which flourished during Fascism (Melis 1996, p. 191; see also p. 107). Special laws introduced the municipalization of public services, such as gas, electricity, and transports, and financed a broad programme of public works both in agriculture and in building the infrastructure of a modern economy in southern underdeveloped areas such as Apulia, Basilicata, Sardinia/Sardegna and Naples. In Benedetto Croce's words, these were "the years in which the idea of a liberal regime was most fully realized" (Croce 1929, p. 214). In contrast, for some, Giolitti was *il Ministro della Malavita* (the Minister of the Underworld) and his was a period of moral decay, political corruption, and crisis of the legitimacy of the state (Salvemini 1910), which paved the way for the Fascist regime (Salvemini 1945). The controversial evaluation of current historiography over the Giolittian era (for a review, see Carter 2011) applies particularly to social and health-related policy. Under Giolitti's tutelage, Italy began to join the path of social legislation that most other countries had previously established. For example, laws were enacted protecting working women and children (1902) and introducing the *Cassa di maternità* (maternity fund) in 1910; instituting the Supreme Council of Labor (1902) and a labour Inspectorate to enforce labour legislation (1912); improving injury compensation and old age insurance, first introduced in 1898 by Di Rudini's and Pelloux's conservative governments. However, while male voting rights made Italian males political citizens, only a few selected groups of them became also "social citizens", invested with statutory rights to some protection against the problems associated with industrialization and urbanization.

Table 1.2 "Social" expenditures by source, selected years (million lire)

	1873	*1880*	*1900*	*1912*
State	1	6	10	59
Municipalities	35	53	78	145
Opere pie	54	64	83	85
Società di Mutuo Soccorso[a]	1.1	2.4	3.2	n.a.[b]
Per cent GDP	1.3	1.6	1.9	2.0

Source: Battilani 2000, p. 645

[a]Mutual Funds

[b]n.a: not available

The great innovation of the Giolittian era in social legislation was that the voluntary principle, which had been ruling the social policy of the liberal governments of the right and of the left, was gradually supplanted by compulsory social insurance. The transition *dalla volontarietà all'obbligo* (from voluntary to compulsion) (Cherubini and Piva 1998) was a messy process amid competing forces which explain the limited spread of social legislation and the fragmented nature of its coverage. During this period, the Italian welfare system acquired its distinctive character of a selective, fragmented, and particularistic scheme. In Giolitti's "insurer state" (Quine 2002, p. 67), insurance programmes resisted the notion of universal obligation marginalizing peasants and selectively covering specific categories of industrial workers of strategic sectors, piggybacking on existing voluntary and self-help institutions. State financial contribution was limited to a bare minimum and kept what we may call state "social expenditures" at around 1 per cent of the Gross National Product (Table 1.2) (Battilani 2000). This constrained coverage and kept benefits meagre. For example, Giovanna Procacci has shown that workers' insurance against occupational injuries covered 5 per cent of the labour force in 1900, grew to 7 per cent in 1903 and to 11 per cent in 1915. Enrolment in *Cassa Nazionale Pensioni* (National Pension Fund) expanded from an estimated 1 per cent of the labour force in 1905 to a mere 2 per cent in 1915, compared to 55 and 51 per cent in Germany (Procacci 2008). Most importantly, social insurance and health policies developed in separate spheres. Industrial accidents and old age were prioritized over sickness insurance, which Giolitti defined in 1913 "a remote ideal…very far away from our times". Separation between social and health legislation meant that working men and women remained excluded from statutory medical and hospital care, benefitting only the poor as with the Crispi public health and social legislation of the late 1880s. Under Giolitti, mutual funds and private charity rather than public welfare continued to provide workers' protection against sickness.

Although a number of health laws were passed under Giolitti's tutelage, they addressed public health in general (1904) and focused on pellagra (1902) and malaria (1901, 1903, and 1907) to be combined in an integrated Code (*Testo Unico*) in 1907, mental hospitals (1904) *Opere pie* and *Ordine dei Medici* (Order of Doctors) in 1910. However, in stark contrast with the grandiose initiatives of his archenemy Francesco Crispi, Giovanni Giolitti relied instead on small adjustments pursued with extreme pragmatism. According to Ernesto Ragionieri, Giolitti saw legislating as a process to be initiated only when intervention could not be further avoided, under the principle "to always concede with the most sluggishness and wariness, and only when further delay is impossible" (Ragionieri 1976, p. 1870). Moreover, he thought that what was needed was not "a first step" in the right direction keeping to "the absolute minimum that is presently possible" (cited in Quine 2002, p. 227). Giolitti's revision of Crispi's 1888 law on public health is a model of this style of legislation. The law's full title was "*Modificazioni e aggiunte alle disposizioni vigenti intorno alla assistenza sanitaria, alla vigilanza igienica ed all'igiene degli abitanti dei comuni del Regno*". That the aim of the amend bill was to make "*modificazioni*" (modifications) and "*aggiunte*" (additions) to "*disposizioni vigenti*" (current legislation) was as telling as qualifying the beneficiaries "the residents of the Kingdom'communes" instead of the Italian citizens. It made explicit its aims at introducing marginal changes while leaving the general design of the system intact. As the Commission in charge of the law was composed in the main by medical doctors, chaired by Angelo Celli, a famous hygienist and malariologist of the Rome University, and had as its speaker, Giuseppe Sanarelli, a medical professor of the same University, ten of its seventeen articles were devoted to fine-tuning the function and status of medical doctors within the system designed by Francesco Crispi in 1888. Key "modifications" focused on *ufficiali sanitari* (medical officers), who were entrusted with an autonomous position from *medici condotti*. They never had the wide authority over the organization of local health services that medical officers of health enjoyed in England, and, for the first time, public health and sanitary functions were split from clinical care, a point that had also been raised and rejected in discussing Crispi's law. *Ufficiali sanitari* were recognized as state officials and reported to *medici provinciali* but were still hired and paid by municipalities. Communes were for the first time allowed to combine resources in *consorzi* (associations) to hire *ufficiali sanitari*, but this did not help much to increase their diffusion, particularly in the south (Cea 2019, p 211). The most significant "addition" was the free provision of drugs for the poor, which were, as usual, to be paid by municipalities, but only if they had not already been provided by charities. Similarly, Giolitti's "reform" of the *Opere pie* was a modest administrative adjustment with the goal of controlling their finances. Additions were the creation of the *Consiglio Superiore di Beneficenza* and an Inspectorate to strengthen state control of charities' budgets and accounts. When pressed by a socialist deputy to push Crispi's secularization project further and to make "public beneficence… a service provided by the state", Giolitti made clear the logic of his reformist attitude, claiming that the Italian state was "too poor to be beneficent". As Maria Quine has aptly noted, Giolitti's odd response in a decade of budget surpluses "say more about him than it does about the possibility for reform" (Quine 2002, p. 62).

Alexander De Grand famously explained Giolitti's strategy of policymaking with the metaphor of the tailor who, in making a suit for a hunchback, must cut its cloth to fit the anatomy of the client (De Grand 2001). Prudence and fiscal conservatism may explain why Giolitti chose not to embark upon a course of sweeping reforms, but to tinker with existing legislation. The lack of an Italian reformist culture also had an impact. The outburst of social legislation, which laid the foundations of the welfare states in France and Britain at the turn of the century, was deeply rooted in ideologies of social reform. In Britain, Lloyd George and Asquith's "radical" or "new" liberalism refashioned traditional, laissez-faire liberalism to cope with the new social problems of the industrial society (Freeden 1978). In France, solidarism ("the official doctrine of the Third Republic"—Hayward 1961) emphasized the need for collective, "organic" solidarity as a response to mutual dependence and took state social insurance the surest and simplest means of collective solidarity. While medico-social reforms in these countries included compulsory sickness insurance schemes, Italy still relied in the main on charity, private beneficence and municipal expenditures. Giolitti's social legislation also confused and divided the emergent workers' organizations (Merli 1972). Trade unions gave their support to the concept of compulsory insurance only lately, and the Socialist party was ambivalent towards social reforms. Socialists fluctuated between a reformist and a revolutionary strategy that provoked intense in-fighting between the directorate of the party and the parliamentary group (Di Scala 1980). In 1897, the Congress of the Socialist Party in Bologna approved a *programma minimo* (minimal programme) of social reforms, which seemed to make collaboration between parliamentary socialists and liberal progressives possible. The Rome Congress three years later adopted instead a revolutionary, maximalist programme. The Lybian war of 1911 produced a backlash against the reformists, who were defeated at the congress of Reggio Emilia in 1912, when Leonida Bissolati and Ivanoe Bonomi were expelled. Both later became ministers; in 1913 Angiolo Cabrini, also a member of the reformist wing, ran a campaign for the adoption of a German-style compulsory scheme of insurance against accidents, sickness, maternity, disability, and old age, funded by the tripartite contributions of state, workers, and employers (Cabrini 1913).

In a period of great advancements of medical science and transformation of clinical practice, Giolittian health legislation grew by accretion through the addition of small and disjointed parts. Angelo Celli, Giolitti's most acute critic, drew a lucid account of this legislation. In one of his last parliamentary speeches, he lamented that "apart some special legislation on malaria, pellagra and rice growing … our sanitary code has remained the same as before, and to stand still implies regress, particularly with respect to other civil nations" (Celli Chamber March 9, 1912). This is far from crediting Giolitti's legislation with the transformation of the old charitable order into a modern system of assistance. Its impact seems instead "a particularistic blend of the old and the new, the traditional and the modern, the religious and the secular, the public and the private" (Quine 2002, p. XIII). Both domiciliary care and the ever-expanding hospital care continued to be limited to the poor and financed by charities and municipalities. In times of economic growth and expanding budgets, which lasted until the Libyan war, Giolitti's fiscal conservatism kept state financial intervention at a bare minimum against the scourges of malaria, pellagra, and tuberculosis, and exposed the nation to the ominous event of a belated cholera epidemic in the year of its Jubilee, which are considered in the next chapter. This may provide some additional arguments why Giolitti's sluggish, cautious, and wary commitment to reform has been famously dubbed ("with exaggerated severity" according to Sassoon—Sassoon 2019, p. 384) "reformism without reform" (Ragionieri 1976, p. 1870).

References

Anonymous (1880) Il Congresso di Beneficenza ed i pericoli delle Opere Pie. La Civiltà Cattolica, 31, pp. 5–19

Anonymous (1887) The hygienic renascence in Italy. BMJ, 2, pp. 728–729

Anonymous (1889) Public health legislation in Italy. BMJ, 1 (1475), pp. 790–791

Anonymous (1892) Direzione generale della Statistica. Statistica delle cause di morte per l'anno 1890. Giornale della Reale Società Italiana di Igiene, 14, pp. 78–86.

Battilani, P. (2000) I protagonisti dello Stato sociale prima e dopo la legge Crispi. In: Zamagni, V., Ed. Povertà' e innovazioni istituzionali in Italia. Bologna, Il Mulino, pp. 639–670.

Benini, R. (1911) La demografia italiana nell'ultimo cinquantennio. Cinquant' Anni di Storia Italiana. Vol.1, Milano, Hoepli, pp. 1–72

Bizzozero, G. (1900) L'igiene pubblica in Italia I. Nuova Antologia,86, pp. 599–612

Borghi, L. (2015) Il medico di Roma. Vita, morte e miracoli di Guido Baccelli (1830–1916). Roma, Armando Editore

Cabrini, A. (1913) La legislazione sociale (1859–1913). Roma, Bontempelli

Carter, N. (2011) Rethinking the Italian Liberal State. Bulletin of Italian Politics, 3 (2), pp. 225–245

Castiglioni, P. (1866) Relazione generale sull'andamento dell'Associazione Medica Italiana e sull'operato della Commissione esecutiva e della sua Presidenza nel triennnio 1863–1866, del dottore Pietro Castiglioni, vicepresidente. Firenze, Annali di Medicina Pubblica

Cea, R. (2019) Il governo della salute nell'Italia liberale. Stato, igiene e politiche sanitarie. Milano, F. Angeli

Celli, A. (1896) Sconforti e speranze d'Igiene sociale. Discorso letto il 5 Novembre 1895 in occasione della solennne inaugurazione degli studi. Annuario della Università degli Studi di Roma.Roma,Tip. Flli Pallotta

Cherubini, A., Piva, I. (1998) Dalla libertà all'obbligo. La previdenza sociale fra Giolitti e Mussolini. Milano, F. Angeli

Contento, A. (1902) Il fenomeno dell'urbanismo secondo i risultati dell'ultimo censimento italiano. Giornale degli Economisti, 25, pp. 207–222

Croce, B. (1929) A History of Italy, 1871–1915, Oxford, Clarendon Press

Davis, J.A. Ed. (2000) Italy in the Nineteenth century. Oxford, Oxford University Press

De Grand, A. J. (2001) The hunchback's tailor: Giovanni Giolitti and Liberal Italy from the challenge of mass politics to the rise of fascism, 1882–1922. Westport, Praeger

Della Peruta, F. (1980) Sanità pubblica e legislazione sanitaria dall'Unità a Crispi. Studi Storici, 21, pp. 713–759

Di Scala, S. (1980) Dilemmas in Italian Socialism. The politics of Filippo Turati. Amherst, University of Massachusetts Press
Duggan, C. (2002) Francesco Crispi. From nation to nationalism. Oxford, Oxford University Press
Favero, G. (2001) Le misure del Regno. Direzione di statistica e municipi nell'Italia liberale. Padova, Il Poligrafo
Fenoaltea, S. (2011) The reinterpretation of Italian economic history: from unification to the Great War. Cambridge, Cambridge University Press.
Franchetti, L. (1877) La Sicilia nel 1876. Firenze, Tip. Barbera
Frascani, P. (1980) Medicina e statistica nella formazione del sistema sanitario italiano: l'Inchiesta del 1885. Quaderni Storici, 45, pp. 942–965
Freeden, M. (1978) The new liberalism: an ideology of social reform. London, Oxford University Press
Gabelli, A. (1890) Il progetto di legge sulle istituzioni pubbliche di beneficenza. Nuova Antologia. 25, pp. 245–274
Giovannini, C. (1996) Risanare la città. L'utopia igienista di fine Ottocento. Milano, F. Angeli
Hanley, J.G. (2016) Healthy boundaries. Property, law and public health in England and Wales, 1815–1872. Rochester, Rochester University Press
Hayward, J.E.S. (1961) The official social philosophy of the French Third Republic: Léon Bourgeois and Solidarism. International Review of Social History, 6 (1), pp. 19–48
King, B., Okey, T. (1901), Italy today. London, Nisbet
Imperatori, U. (1891a) La nuova politica sanitaria in Italia I-II. Il Giornale degli Economisti, 2, pp. 246–290
Imperatori, U. (1891b) La nuova politica sanitaria III. Le finanze comunali e la esecuzione della legge. Il Giornale degli Economisti, 3, pp. 371–410
Imperatori, U. (1891c) L'ingerenza dello Stato IV. Il Giornale degli Economisti, 4, pp. 579–615
ISTAT (2011) L'Italia in 150 anni. Sommario di statistiche storiche, 1861–2010. Roma
Jacini, S. (1885) Relazione finale sui risultati dell'Inchiesta. In Atti della Giunta per la Inchiesta Agraria e sulle condizioni della classe agricola. Vol. XV. Roma, Forzani & C. Tip. Senato
Levra, U. (1982) Il colpo di stato della borghesia, 1896–1900. Milano, Feltrinelli
Malatesta, M. (2011) Professional men, professional women. The European professions from the 19th century until today. Los Angeles, SAGE
Marucco, D. (1980) Mutualismo e sistema politico. Il caso italiano, 1862–1904. Milano, Franco Angeli
MAIC—Ministero dell'Agricoltura, Industria e Commercio. Direzione di Statistica (1880) Ordinamento della Statistica delle Cause di morte. Annali di Statistica, Roma, Tip. Eredi Botta, pp. 187–196

MAIC—Ministero dell'Agricoltura, Industria e Commercio. Direzione di Statistica (1886a) Annuario Statistico. Annali di Statistica, Roma, Tip. Eredi Botta
MAIC—Ministero dell'Agricoltura, Industria e Commercio. Direzione di Statistica (1886b). Le condizioni sanitarie dei comuni del Regno. Parte Generale. Annali di Statistica, Roma, Tip. Eredi Botta
Melis, G. (1996) Storia dell'amministrazione italiana, 1861–1993. Bologna, Il Mulino
Merli, S. (1972) Proletariato di fabbrica e capitalismo industriale. Firenze, La Nuova Italia
Mortara, G. (1908) Le popolazioni delle grandi città italiane al principio del secolo ventesimo. Torino, UTET
Nitti, F.S. (1892) Poor relief in Italy. Economic Review, 2(1), pp. 1–24
Ognibeni, G. (1982), Legislazione ed organizzazione sanitaria nella seconda metà dell'Ottocento. In: Betri, M.L., Gigli Marchetti A., Eds. Salute e classi lavoratrici in Italia dall'Unità al Fascismo. Milano, F. Angeli, pp. 583–603
Pagliani, L. (1889) L'Igienista nello Stato moderno. La Rivista della Beneficenza Pubblica e delle Istituzioni di Previdenza, 16, pp. 11–18
Pagliani, L. (1890) Relazione intorno all'ordinamento della direzione della sanità pubblica ed agli atti da essa compiuti dal 1° Luglio 1887 dal 31 dicembre 1889. Giornale della Società Italiana d'Igiene, 2–3, pp. 73–115.
Pagliani, L. (1901) Giulio Bizzozero: commemorazione fatta alla Società Piemontese d'Igiene il 27 Aprile 1901. Torino, Stab. Flli Pozzo
Pancaldi, G. (1983) I Congressi degli Scienziati Italiani nell'età del Positivismo. Bologna, CLEUB
Pareto, V. (1891) Statistica dolorosa. Il Secolo, 22–23 April
Patriarca, S. (1994) Statistical nation building and the consolidation of regions in Italy. Social Science History, 18, pp. 359–376
Patriarca, S. (1996) Numbers and Nationhood. Writing statistics in nineteenth-century Italy. Cambridge, Cambridge University Press
Pollard, J. (2011) A State within a State: the role of the Church in two Italian political transitions. Modern Italy, 16, pp. 449–459
Pollard, J. (2014) Catholicism in modern Italy. Religion, society and politics since 1861. London, Routledge
Procacci, G. (2008) Le politiche di intervento sociale in Italia tra fine Ottocento e Prima guerra mondiale. Alcune osservazioni comparative. Economia & Lavoro, 42, pp. 17–43
Quine, M.S. (2002) Italy's social revolution. Charity and welfare from liberalism to fascism. Basingstoke, Palgrave
Ragionieri, E. (1976) La storia politica e sociale. In Storia d'Italia, vol. IV. Dall'Unità ad oggi. Torino, Einaudi
Rosenberg, C. (1987) The care of strangers. The rise of America's hospital system. New York, Basic Books

Salvemini, G. (1910) Il Ministro della malavita. Notizie e documenti sulle elezioni giolittiane nell'Italia meridionale. Firenze, Ed. La Voce

Salvemini, G. (1945) Introductory Essay. In: Salomone A.W. Italian democracy in the making. The political scene in the Giolittian era, 1900–1914. Philadelphia, University of Pennsylvania Press (XIII–XXIII)

Sassoon, D. (2019). The anxious triumph. A global history of capitalism, 1860–1914. London, Allen Lane

Scotti, G. (1889) La riforma della legge sulle Opere Pie. La Rivista della Beneficenza Pubblica e delle Istituzioni di Previdenza, 16, pp. 3–10

Serao, M. (1884) Il ventre di Napoli. Milano, Treves

Silei, G. (2003) Lo Stato sociale in Italia. Storia e documenti. Vol. 1. Dall'Unità al Fascismo. Manduria, Lacaita

Somogy, S. (1967) La mortalità nei primi cinque anni di vita in Italia, 1863–1962. Palermo, Ingrana

Sori, E. (1984) Malattia e demografia. In: Della Peruta F., Ed. Storia d'Italia. Annali 7. Malattia e Medicina. Torino, Einaudi, pp. 541–588

Sormani, G. (1881) Geografia nosologica dell'Italia. Annali di Statistica, Roma, Tip. Eredi Botta

Sonnino, S. (1897) Torniamo allo Statuto. Nuova Antologia, 151, pp. 9–28

Strambio, G. (1862) Sull'organizzazione sanitaria in Italia. Il Politecnico, 14, 245–307

Toniolo, G. (2013) The Oxford Hanbook of the Italian economy since unification. Oxford, Oxford University Press

Villari, P. (1878) Le Lettere Meridionali ed altri scritti sulla questione meridionale in Italia. Firenze, Le Monnier

Villari, P. (1890) La riforma della beneficenza. Nuova Antologia, 114, pp. 5–40

Villari, P. (1893) Dove andiamo? Nuova Antologia, 132, pp. 5–24

White Mario, J. (1890) Scritti e Discorsi di Agostino Bertani. Relazione all'onorevole Ministro dell'Interno che accompagna il Codice per la Pubblica Igiene, compilato e proposto dal deputato dottore Agostino Bertani, 1885. Firenze, Tip. Barbera, pp. 349–370

White Mario, J. (1895) The poor in great cities. Their problems and what is doing to solve them. New York, Scribner's Sons

Wohl A. S. (1983) Endangered lives. Public health in Victorian Britain. Cambridge, Harvard University Press

Woods, R. (2003) Urban-Rural mortality differentials. An unresolved debate. Population Development Review, 29, pp. 29–46

Woolf, S. (2000) The "transformation" of charity in Italy, 18th–19th centuries. In: Zamagni, V., Ed. Povertà e innovazioni istituzionali in Italia. Bologna, Il Mulino, pp. 421–440

Zamagni, V. (1993) The economic history of Italy, 1860–1990: recovery after decline. Oxford, Oxford University Press

CHAPTER 2

Health and Healthcare in the Liberal State

In Italy, mortality declined steadily from the 1880s to the First World War, largely because of a decrease in deaths from infectious diseases, particularly in the younger age groups and in the big cities of the northern part of the country. The decline has traditionally been attributed to improvements in the standard of living, especially better diet and housing conditions, and to the diffusion of new public health practices and sanitary works, particularly the provision of central water supplies and sewage systems. Moreover, the increase in medical knowledge and technical innovation brought about by the bacteriological revolution at the turn of the century significantly changed medical practice in hospitals and the community. These innovations affected some conditions more than others and were of variable import in different geographic areas and population groups. The manifold public health practices advocated by the public health movement and/or prescribed by public health legislation described in the previous chapter were implemented very differently, because of the lack of human capacity or financial resources, political conflict at the national or municipal level, resistance from interest groups or the general population.

This chapter examines state and popular responses to a set of "exemplary diseases" (Kunitz 1988) broadly characteristics of rural and urban areas, including cholera, malaria, and pellagra. Malaria and pellagra represent two highly prevalent conditions, whose history is strictly associated with the evolving political economy of agriculture, including agricultural intensification and peasant mobilization. The four waves of cholera

F. Taroni, *Health and Healthcare Policy in Italy since 1861*,
https://doi.org/10.1007/978-3-030-88731-5_2

epidemics that hit Italy after its unification were relatively minor contributors to overall deaths, but their dramatic appearances had enormous political impact, "forcing" a number of sanitary measures and, among all, the passing of Crispi's law on public health. Tuberculosis was the most frequent cause of death at about 50,000 deaths per year (Fig. 2.1) but will be considered in Chap. 4. Despite high frequency and number of deaths, and the increased knowledge over the disease, the anti-tubercular policy of the Giolittian era was "a gigantic episode of collective neglect" (Detti 1982, p. 29), and significant interventions had to await the compulsory insurance introduced by the fascist regime. This chapter also explores the increasing "medicalization" of the Italian society, considering the changes in doctors and their practice at the community level, including the emerging specialization of *medicina del lavoro* (occupational medicine). A special focus is on the changing social function of Italian hospitals and their increasing "medicalization" and specialization.

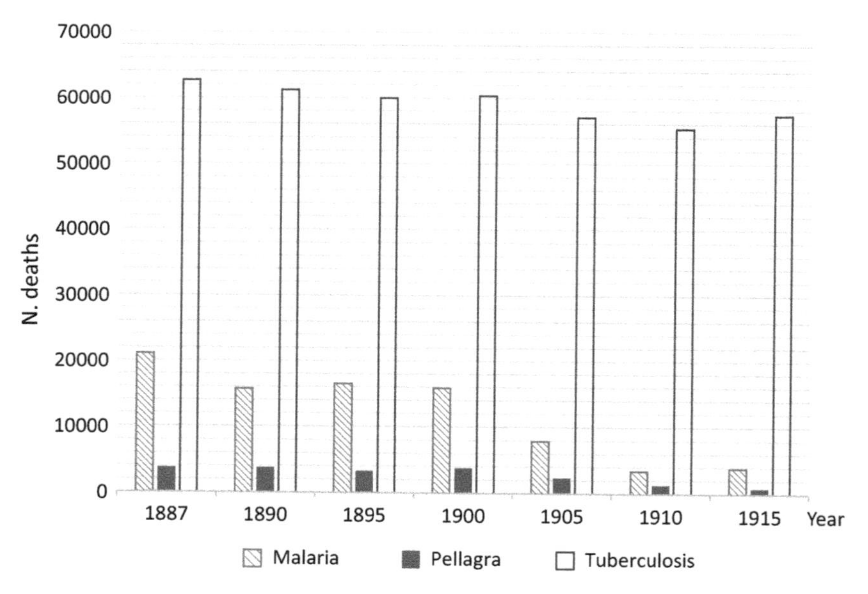

Fig. 2.1 Number of deaths from malaria, pellagra, and tuberculosis, 1887–1915, selected years

Cholera, the Nineteenth-Century Disease

Cholera is "the classic epidemic disease of the nineteenth century" (Evans 1988, p. 125) and "a tool for social and economic analysis" of the infected communities (Rosenberg 1966, p. 454). Margaret Pelling had rightly observed that, when compared to endemic conditions such as fevers, typhoid, and tuberculosis, cholera epidemics had a relatively minor impact on the health of the population in terms of deaths and the number of persons infected (Pelling 1978, p. 3). However, the incoming cholera pandemics prompted a series of International Sanitary Conferences, which, on the one hand literally "unified the globe" (Huber 2006) and, on the other, intensified commercial competition between France, England, and Germany. In Italy, the responses to the four main waves of the Liberal era exposed the changing political and social cultures of national and local authorities, tried their capacity, and provided a sanitary stress test of previous interventions. Moreover, cholera epidemics were a catalyst in developing sanitary movements and "forcing" broader hygienic reforms (but see also Hamlin 2009).

In the nineteenth century, Italy was visited six times by cholera, one wave for each major pandemic, except the first that did not reach Europe (Forti Messina 1984; Tognotti 2000). Between unification and the First World War Italy experienced four waves of cholera beginning in 1865, 1884, 1893, and 1910. Table 2.1 reports the number of deaths for each epidemic wave from two of the most authoritative sources, Del Panta (1980) and Forti Messina (1984). They independently assembled data on cholera-associated deaths from various sources, including official statistics

Table 2.1 Number of deaths from the main waves of cholera epidemics in Italy, 1861–1911

Years	*Forti Messina*	*Del Panta*
1835–1837	236,473	146,383
1849	17,211	13,359
1854–1855	248,514	118,530
1865–1877	160,547	160,547
1884–1887	33,875	21,958
1892–1895	3037	3040
1910–1911	6950	–
Total	700,352	463,817

Sources: Del Panta (1980) and Forti Messina (1984)

(MAIC 1886) and contemporary reports (see, e.g., Corradi 1876, for epidemics prior to unification; Pagliani 1894; Raseri 1895; Celli 1896). Estimates of the cumulative number of cholera deaths in Italy from unification to the 1910–1911 epidemic range from about half a million to slightly over 700,000 (Alfani 2014). In terms of sheer mortality, only the most severe 1865–1867 epidemic compares with tuberculosis, which caused about 50,000 deaths a year. The number of people infected declined in each wave, possibly the effect of the increased capacity to contain the spread of the epidemics, while lethality remained stable at about 50 or 60 per cent.

The first cholera epidemic was the most severe and ravaged several areas of the country beginning in 1865, reaching its peak in 1867, when, according to Angelo Celli, it led to 128,073 of the total of 160,045 deaths (Celli 1896). The epidemic hit the countryside and small towns more than the largest cities and concentrated in the southern regions more than in the north. However, it also hit the large cities of Naples and Palermo, possibly through the Army sent from the garrison of Naples to suppress the Sicilian September unrests (Riall 1995). The state applied a strategy of heavy-handed military intervention through harsh cordoni sanitari, aimed more at controlling social riots than containing the epidemic (Dickie, 1999; for a praising contemporary report on the Italian army see also De Amicis 1869). As it was concentrated in the southern part of the country, this epidemic was easily explained away as the after-effect of the Bourbon's sanitary neglect and backwardness, assuming that the new Kingdom could not be held accountable for the filth and poor urban environment that had bred cholera (Maggiori-Perni 1894).

In contrast, the 1884–1887 cholera outbreak was a truly national event, which brought significant changes both in national containment policies and in the political and administrative life of Naples, its epicentre. This helped create a political space for public health and sanitation, which transformed into matters of "high politics" for national politicians and bureaucrats and opened the way to the approval of Crispi and Pagliani's public health law of 1888. When cholera first hit Italy in 1884, Robert Koch had just isolated *Vibrio cholerae* in his mission to Alexandria and Calcutta, but the discovery was still questioned and was just beginning to have an impact on the choice of intervention strategies (Ogawa 2000). The cholera epidemic had broken out in Italy following seasonal agricultural workers escaping the infection in Provence (France) through the Alps. In Parliament, representative Guido Baccelli, professor of clinical medicine at the University of Rome, pestered Prime minister Agostino Depretis to send the Army to "seal the Alps" with sanitary cordons, strict quarantine, and fumigation (Baccelli Chamber 27–28 June 1884). The parliamentary debate spilled over into the intellectual circles with several articles in the well-read magazine *Nuova Antologia* and divided the Society of Hygiene. *Il Giornale della Reale Società Italiana di Igiene* reported radical disagreements between the schools of Turin and Milan, the two strongholds of the hygienic movement in Italy. While Carlo Zucchi from Milan found the containment strategy advocated by Baccelli "coherent with the tradition of the Italian medical school and consonant with modern science" (Zucchi 1884), Luigi Pagliani and the Turin School insisted instead on adopting what was later called "the English system of prevention" (Thorne-Thorne 1887) based on local sanitation and "active surveillance by visits and inspections" in order to promptly "stamp-out" early cases (see also Hardy 1993).

Angelo Mosso, a senator and professor at the University of Turin, brought the controversy into public view with a paper published by *Nuova Antologia*, vigorously attacking the government's policy. His paper dubbed Baccelli's idea of sealing the Alps "a medieval concept", which Pasteur, Virchow, and Pettenkofer opposed, and "a failure of Italian medicine" (Mosso 1884, p. 330). One year later cholera had invaded Italy through several routes, and an anonymous "former official of the Ministry of the Interior" acknowledged the failure of coercive measures in halting the epidemic but only because their implementation had not been strict enough (Anonymous 1885). Pagliani's attack on coercive measures as "useless" and "not in accordance with latest scientific findings" (Pagliani 1886) came out when he oversaw managing the epidemic and had already overturned the old strategy of sanitary cordons and quarantine with sanitation and prompt detection and isolation of early cases. Luigi Bizzozero, the saviour of the Turin School, explained the turnaround in Italian policy at the congress of the Italian Medical Association in Pavia, a major centre of the Italian medical elite, making clear that he was speaking on behalf of Prime Minister Francesco Crispi, who had just come to power after Depretis' sudden death. His claim that coercive measures had led to "maximum anarchy", aroused "an arsenal of fear", and provided "a deceptive safety from misunderstood science" won the praise of the anonymous English correspondent for the *British Medical Journal* who welcomed "the hygienic renascence in Italy" (Anonymous 1887).

In Italy, however, politics and public opinion were still divided, and the demand for protection against the exogenous penetration of cholera led to unrest no less than the imposition of traditional restrictions. Some newspapers opposed quarantine, lamenting its ineffectiveness against cholera and the blockage of trade and tourism (see, e.g., Il Messaggero from Rome, August 1884). The Giornale di Sicilia demanded, instead, the blockage of ships from Naples and argued forcefully against "the theory of freedom…the easiest, for it does nothing" (Maggiori-Perni 1894, p. 504). To try to put an end to domestic quarrels, Italy called an International Sanitary Conference, which was held in Rome in September 1885. The Conference however failed to provide a clearinghouse for alternative scientific interpretations and conflicting measures (Howard-Jones 1975). Scientific and diplomatic delegations maintained their opposing views on containment or quarantine, mostly because of the competing commercial and imperial interests between the leading countries of Britain, France, and Germany.

A wide consensus existed about the dire conditions in the city of Naples, the epicentre of the cholera epidemic in Italy, where two-thirds of the cases and half of total deaths were concentrated. Jessie White Mario (1895) and Matilde Serao (1884) had widely noted the conditions of those living in the *fondaci* of the lower city, which Axel Munthe later stigmatized as "the most ghastly human dwellings on the face of the earth" (Munthe 1909). Naples, the ancient capital of the Bourbons' Kingdom of the Two Sicilies was still the largest Italian metropolis and one of the major cities of Europe, a focal point for the southern part of Italy, and an increasingly important seaport for transatlantic trade and emigration. The magnitude of the hygienic crisis and the city's pivotal role in the social and political economy of the South and of the whole country put the health of the whole nation at risk. It also became a threat to the legitimacy of the Liberal rule, which the Jesuit review *Civiltà Cattolica* was quick to emphasize. For the Catholic Church the cholera epidemic was *flagello e maestro* (scourge and teacher), a scourge sent by God as a vengeance for the sins of the Kingdom and a penance to redeem Italians from "the sordid and materialistic" ruling of the "masonic sect", that is, the liberal elite (Anonymous 1884).

A turning point in the fight between the Church and the state was the visit to Naples of King Umberto I and his Court, followed by Prime Minister Depretis with three Ministers, who "fortuitously" met Guglielmo di Sanfelice, Cardinal and Archbishop of Naples. All walked the most stricken quarters of the city, an "act of courage", and a sign of settlement between the state and the Church that the press duly noted (Brice 2002, p. 73). Pushed by the King's sense of urgency that "Italy must redeem Naples at any cost!" which Depretis' famous imperative translated into *dobbiamo sventrare la città!* (we must disembowel the city), and with the Pope's blessing, an ambitious and expensive plan quickly followed (Logan 2002). The Parliament passed the *Legge per Napoli* by acclamation in record time as a sign of national solidarity and debt of gratitude for the role played by the "neglected city" in the process of unification. The law financed with 100 million liras in transfers and loans a plan of *risanamento* (sanitizing) and *miglioramento* (improvement) of the city following the lines recently pioneered by Baron Haussmann in Paris (see, e.g., Gandy 1999). The plan included, among several other things, the clearing of the slums and the building of new apartments, opening a central boulevard with intersecting avenues to act as a "great urban lung" for the city, and improving the sewage and water system using the fresh water of the Serino River. The law established a model for a national policy of direct state intervention in public works, which was quickly expanded to other communes (Giovannini 1996), and a strategy of financing structural shortcomings of the southern territories through *leggi speciali* (special laws), which was perfected in the Giolittian period by Francesco S. Nitti with his plan for industrializing Naples (Melis 1993).

The implementation of the massive programme of *risanamento* placed public works at the centre of the political and administrative life of the city and tested municipal capabilities. Public funds were managed by a private monopolistic contractor, *the Società per il Risanamento*, a banking conglomerate put together with great difficulty by Francesco Crispi in 1887, which started its works only in 1889 (Giovannini 1996). The results were far from satisfying as synthetized by Paolo Villari's famous claim *era meglio il colera!* (Cholera was better!—Villari 1890). People evicted from the demolished *fondaci* who could not afford the new apartments' rents were displaced by middle-class tenants and relocated to remaining slums. The programme also tapped a rich vein of patronage, clientelism, and sheer corruption for the infiltration of *camorra*. The municipality was put under the administration of an external commissioner and then subject to the Saredo's Royal Commission of Inquiry, documenting mismanagement of funds and control of the city by the "high" or "political *camorra*" (Snowden 1995, p. 206). The many failures of the ambitious project made Naples easy prey for the next epidemics.

When in 1910 cholera struck again Italy, and Naples in particular, it was a disease of the past. Koch's etiologic theory was universally acknowledged, and optimal measures of prevention and control were fully accepted and incorporated into the Paris International Convention of 1903 to which Italy was a signatory. Tullio Rossi Doria's statement that "*il colera è di chi lo vuole*" (cholera is for those who want it) summarized the state of the art. As "the rich can easily avoid it", cholera was now only "the disease of the poor and the hungry, the ignorant and the superstitious, the symbol of the backwardness of a country" (Rossi Doria 1910). Italy was also different. It was going through a process of rapid industrialization and economic development and was preparing the celebrations of its *Giubileo* (Jubilee), the fiftieth anniversary of its unification, widely seen as the opportunity for taking its long-deserved seat at the table of the great powers of Europe. Naples was still among the larger Italian and European cities and one of the biggest and busiest transoceanic seaports, the main hub of the massive process of emigration whose hard currencies' remittances enriched Italian coffers and fuelled the Neapolitan economy (Barbagallo 2015). The public health infrastructure had also improved, and the *Direzione generale di sanità* had been restored by Giovanni Giolitti to its original strength and power after the "dark years" of its downgrading under Di Rudini. The westward progression of the pandemic was carefully watched over, and in 1905 its arrival in Italy was declared "not just probable but imminent", and a plan to counter the incoming epidemic was developed (Snowden 1991, p. 76). However, when cholera entered the Italian territory, probably through one of the long-distance fishermen sailing from the small ports of the lower Adriatic Sea, municipal public health services proved incapable of promptly "stamping-out the first few cases" as theory and good practice of containment suggested. Infection spread rapidly heading for Naples through the backward territories of Apulia and Calabria, which had changed little since unification. Despite Crispi and Giolitti's reforms, public health and sanitation were seriously wanting, and "medical attention was infrequent, unwelcome and often inexpert" (Snowden 1991, p. 92). This raised anxiety and uncertainty about whether public works and sanitation had really made Naples resilient to the return of cholera (Tropeano 1912).

As in 1884, the return of cholera to Naples was a serious threat to the city and to the nation as a whole. To the city, the epidemic provided the ultimate sanitary test of risanamento and risked blocking its port and severely damaging commerce and the economy of the city. Cholera also threatened celebration of the *Giubileo della Patria* (the jubilee of the homeland), the opportunity to celebrate Italy's achievements, and damaged the reputation of the nation, which pretended modernity but was instead plagued by the very symbol of backwardness. In a context of patriotic nationalism and pressures from local interests, the national government adopted a policy of denial, hiding or obscuring the real number of cases and deaths. Instructions were given to report only laboratory-confirmed cases of Asiatic cholera. The poor availability of laboratory facilities obviously deflated notifications, and municipal bulletins reported an increasing number of cases of gastroenteritis or diarrhoea, and, in case of swift death, meningitis (Tropeano 1912, p. 152). Professional journals, including the *Giornale della Reale Società di Igiene*, practised self-censorship, obliquely referring to "the disease we are not allowed to name but everybody knows". Parliamentary speeches denied cholera in Italy and particularly in Naples and denounced "the sordid interests" of foreign countries and their *prezzolata* (bribed) press to "disparage Italy" for breeding cholera (Di Scalea Chamber January 30, 1911).

Nevertheless, *The Lancet* of August 12, 1911, reported that "correspondents in the continental press alleged that the true incidence of cholera in Italy is not recorded in the returns published by the Italian government, and that the occurrence of outbreaks is being concealed by the authorities". It was not just a matter of transparency. The crucial issue was the conflict between the national interest in emigration, pressures from the Naples' municipality to keep the port fully functioning, and the 1903 Paris Sanitary Convention that required full disclosure of cases of cholera by the port of departure. The conflict engaged several players with competing interests. With New York as the main destination of the emigrants sailing from Naples, the United States' government was a major player, interested in the full disclosure of the status of the infection in the city of Naples and among the emigrants from all over the south that was granted by the Paris Convention. However, pushing American interests too far in controlling its borders to preserve the health of its population disturbed commerce and risked embarrassing a friendly government. *The New York Times* reported several cases of cholera on board of Italian steamships from Naples. *Public Health Reports*, the official journal of the US Public Health Service, published several alarming but ineffectual accounts, which were anonymous but possibly came from Henry Geddings, director of the medical services of the Naples' US Consulate (see, e.g., *New York Times*, July 18 and 23; Anonymous 1910, 1911).

Local protests in Naples aimed to prevent the adoption of the coercive measures which would necessarily follow from admitting the presence of cholera. Protest took the form of lockouts of firms, closures of shops, and mass demonstrations, and was organized from the top, including the Chamber of Commerce, the mayor, and the municipal council, supported by the moderate local newspaper *Il Mattino*. They differed greatly from spontaneous popular disturbances in the small country towns and villages against coercive measures, which were perceived to break local customs and traditions or restricting local commerce of small items such as mussels, fish, or produce. Rioters in Barletta, Molfetta, Bari, and dozens of other smaller villages in Apulia and Calabria stormed town halls, post offices, and hospitals, and attacked governmental officers, including medical doctors and pharmacists as their supposed agents (Snowden 1991). The most dramatic episode occurred at the end of August 1911 in Verbicaro, a small, remote village in the Calabrian hills. The spread of cholera in the small village was associated with the recent census of the population. A conspiracy theory denounced a Malthusian plot of the local authorities to curb people in excess by getting rid of the poorest part of the population through a "cholera powder" in the town well (Cohn 2017). During Verbicaro's bloody riots, the mayor and a dozen local authorities who had not yet fled the village were killed. The event made front-page headlines of the major national newspapers for weeks. The Turin-based, pro-government newspaper *La Stampa* published an interview with Prime Minister Giovanni Giolitti besides a report with the heading "The savage uprising of Verbicaro", where Giolitti called the event "an episode of mass folly" to be dealt with by "force and well-focused power" (La Stampa 29 August 1911). The *Corriere della Sera* (a traditional political opponent of Giolitti's government) sent the famous reporter Luigi Barzini who under a front-page headline of "*Verbicaro in pieno Medioevo*" (Verbicaro in full Medieval times) ascribed the event "to another and different race, and a different and faraway time" (Corriere della Sera 31 August). In a second report a few days later (4 September) "*Una terra italiana da redimere*" (An Italian land to redeem) Barzini argued for a national mission of the modern and progressive northern part of Italy as the only way to the recovery of the whole country. Barzini insisted on the sense of "otherness" and the "horror of diversity" provoked by disease- and crime-ridden places and populations and identified for the Liberal state a "civilizing mission", understood as a colonial project of the progressive North to the backward South of the nation. This resonated with historically constructed stereotypes (Dickie 1999) and Alfredo Niceforo's theory of the two Italies, where the South was Latin and barbaric and the North Germanic and civilized (Niceforo 1901).

The last cholera epidemic dramatically exposed the gulf separating the northern from the southern part of Italy, which had been made wider and deeper by the recent spurt of industrialization. The same concept had been represented by Giustino Fortunato's metaphor of the *Due Italie* (Two Italies), which had headed the front page of *La Voce* a few months earlier (Fortunato 1911). Less well represented were the wide differences among the "many Souths", which included both Naples and Verbicaro, casting legitimate doubts on the idea of a uniform, homogeneous Southern society with a common set of problems and failings. Internationally, in Angelo Celli's words in Parliament, the 1910–1911 cholera epidemic had "split Europe into two parts", and the "backward and cholera-ridden" part included Italy, with Russia and Turkey. This had put Italy in the embarrassing position of being "object of resentment and fear as a source of danger and a threat of epidemic for the civilized world" (Celli Chamber March 9, 1912). More than disrupting the celebration of its Jubilee, the cholera epidemic reflected Italy's status as "the least of the great powers" (Bosworth 1979) and frustrated the enduring effort of reinstating the country among the European power elite.

The Rise and Demise of Pellagra: A Success Story

Pellagra has had a long history in Italy, as the earliest descriptions of *pelle agra* (rough skin) by the peasants who suffered from it date from the 1770s. However, it was recognized as a national problem only in the second half of the nineteenth century (De Bernardi 1984). From the time of its initial appearance, pellagra was associated with the diffusion of maize cultivation and its consumption as a low-cost dietary staple by peasants, particularly the landless day—labourers, the poorest among them. While the association of pellagra with maize was clearly perceived, causal interpretations differed substantially and entailed different policy implications. Causal theories hypothesized either an intoxication caused by rotten maize and/or infected by a mould (Lombroso 1869) or the insufficient nutritional content of a maize-based diet (Lussana 1872) and aimed either for prohibiting consumption of spoiled maize or for promoting better and more varied nutrition. Since maize was consumed where it was produced, there was a correspondence between the geography of production and consumption and the distribution of the disease. The diffusion of pellagra was limited to the regions of the "pellagra triangle" (Veneto, Lombardy, and Emilia-Romagna), where rapid population growth and the intensification of agriculture led to extensive cultivation of maize which made up about half of national production. The southern part of the country, where maize cultivation was absent, as well as the cities, where the typical diet did not include maize, remained untouched by the disease. General improvement in the economic conditions and the mechanization and modernization of agriculture rapidly reduced the incidence of the disease, which virtually disappeared by the end of the First World War (Whitaker 1992; Brown and Whitaker 1994).

Table 2.2 provides information on the total number of cases and new cases of pellagra, along with their associated deaths. Data on pellagra cases and mortality are only indicative of the general trend since official figures significantly underestimated the number of pellagrins, who had limited access to services and whose symptoms were often confused with other diseases. Moreover, national death statistics for the whole country are available beginning only in 1887, when the peak of the disease had already been reached. A report in 1879 noted 97,855 cases (Miraglia and Sormani 1880, p. 207), slightly less than the 104,067 cases of the first official national survey of 1881 (MAIC 1886). In the 1880s, the prevalence of pellagra started to decline, faster in Lombardy and Emilia and more slowly in Veneto (Sitta 1899). The concomitant rise in the four regions of Piedmont, Tuscany, Umbria, and Marche reflected a new social stage of

Table 2.2 Pellagra: total cases, new cases, and deaths, 1879–1930, selected years

Year	*Total cases*	*New cases*	*Deaths*
1879	97,655	n.a.	
1881	194,067	n.a.	3688
1899	72,603	n.a.	2836
1905	55,029	3018	2357
1910	33,861	2176	1312
1915	n.a.	951	811
1920	n.a.	256	331
1925	1446	103	108
1930	n.a.	69	81

Sources: Miraglia and Sormani Annali di Statistica, 1880; ISTAT, Cause di morte, 1887–1955, Roma, 1958, pp. 74–75

the disease which spread to low-fertility lands where contracts of *mezzadria* (sharecropping) endured. After the turn of the century, the incidence of new cases rapidly declined, and chronic and elderly cases represented the majority of pellagrins who were frequently found in *manicomi* (asylums). In 1910, the prevalence was little more than one-quarter of that in 1881, and by the 1920s pellagra had practically disappeared except for a few scattered cases, reaching a total of 1466 in 1926. Death statistics confirm the trend in morbidity. Mortality rates per 100,000 population between 1887–1891 and 1912–1916 declined from 228 to 60 in Veneto, 152 to 23 in Lombardy, and 118 to 12 in Emilia (Livi-Bacci 1986). Benito Mussolini officially declared the disappearance of pellagra in his famous Ascension Day speech of May 26, 1927. By contrast, pellagra was just reaching its peak in the southern states of the United States, where it had appeared only in the early 1900s and spread so rapidly that was initially regarded as a contagious disease (Roe 1973; Sydenstricker 1958).

Pellagra is a success story in the social history of population health in Italy, but scientific medicine played a small part. Pellagra is a classic example of a disease for which preventive measures were adopted long before its complete understanding and exact cause were identified. For the time span of pellagra's history in Italy, there was no single agreed-upon theory about its aetiology. However, Cesare Lombroso's toxic theory based on the consumption of spoiled maize came to dominate Italian medicine, through Pellagrological local conferences and the journal *Rivista Pellagrologica Italiana*, influenced local policies, and determined national legislation. Lombroso stressed the political and practical expediency of his theory compared with the deficiency concept, which he dubbed "a cruel irony" and "a useless truth" for the poor peasants. "To recommend that peasants not consume maize is tantamount to telling them to become rich; but to prevent them eating some of it when it is spoiled … this is not beyond the limits of the possible" (Lombroso 1869, p. 153). The persuasiveness of Lombroso's remedy to pellagra was also found in the analogy with the dominant bacteriological model of intervention in case of food infection, as with "the prohibition of consuming ham jam infested with trichinella, poisonous mushrooms, and rotten fruits" (Miraglia and Sormani 1880, p. 210).

Liberal governments of the right and of the left did very little for a very long time, while pellagra cases and deaths continued to increase. Initiatives were primarily at the local level through *Commissioni Pellagrologiche*, which followed the spoiled maize theory but took a pragmatic attitude with activities more attuned to the deficiency concept, such as *cucine economiche* (low-cost kitchens) providing soup and bread to the poorest peasants (Sitta 1900). The two early governmental bills prohibiting commerce and consumption of spoiled maize and providing for municipal *essicatoi* (driers) for maize prepared in 1887 failed to reach the floor of the Parliament (De Bernardi 1984). By the time a national law was enacted in 1902, pellagra was already in decline. In the early 1900s new forces were active to spur governmental inertia and overcome landowners' resistance. Italy was experiencing a rise in labour organization and agitation, which had its epicentre in the Po Valley, at the core of the pellagra triangle and was also backed by the militancy of *medici condotti* (Detti 1979).

Giuseppe Zanardelli's government, with Giovanni Giolitti as Minister of the Interior, was more open to social issues and agreed to some state intervention, which was demanded by the wide movement of local *Commissioni* and *Congressi Pellagrologici* and supported by authoritative scientists and influential doctors. Speakers for the law were Nobel laureate Golgi at the Senate and at the Chamber Nicola Badaloni, socialist, *medico condotto*, and member of the *Consiglio Superiore di Sanità*, who talked of the law as "the word of science becoming law". The focus of the law of July 21, 1902, was the Lombroso-inspired battle against *muffe* and *truffe* (moulds and frauds) prohibiting the commerce and consumption of immature, spoiled, and rotten maize. The law sanctioned the import, trade, mill, and use of unripe and spoiled maize and its products; officially recognized the authority of Provincial Pellagrological Committees; and obliged doctors to notify cases of pellagra and municipalities to keep an updated register of pellagrins. The law also gave new impetus to traditional local initiatives more consonant with the dietary concept, recommending municipalities to provide for their registered cases the distribution of salt and *cura alimentare* (nutritional treatment) through soups, economic kitchens, health stations (*pellagrosari*), and hospitalization in specially designed "pellagra wards", as well as promoting the breeding of rabbits to increase protein intake.

The steady decline of pellagra in the following years seemed to vindicate Lombroso's theory even if a host of further empirical studies failed to reproduce the disease. Lombroso himself, who claimed to have isolated the toxin concerned and the germ producing it, proposed several refinements and variants of his toxic theory (Lombroso 1892). Reviewing thirty years of pellagra policies, the undersecretary for agriculture and university professor Giuseppe Sanarelli expressed dissatisfaction with scientific status and legislative assets about pellagra, lamenting that "the disease has not yet started to show a fair decline in Italy" and that "current provisions could not be considered either sufficient or appropriate to stop maize intoxication for good" (Sanarelli 1909). Dissatisfaction was also reflected in the careful consideration of Louis Sambon's disruptive theory that maize had nothing to do with pellagra, which was a protozoal disease caused by the bite of a sand-fly (Sambon 1905; see also Gentilcore 2016). Sambon's theory was examined by a special *Commissione Ministeriale per lo Studio della Pellagra* (Ministerial Commission for the Study of Pellagra), set up to test various causal theories (except the "right" Casimir Funk's deficiency theory—Funk 1912) and their impact on the 1902 law.

The "New etiologies about Pellagra" were the focus of the fifth Italian Pellagrological Congress in 1912, which concluded in favour of Lombroso's theory, although most *medici condotti* were now more attuned to the deficiency theory (De Bernardi 1984, p. 258). In 1922, when pellagra had almost completely disappeared, Gosio and Antonini, two of the most important pallbearers of Lombroso's theory concluded in their paper delivered at the sixth *Congresso Pellagrologico Italiano* that the toxic theory had not been proved right but "preventive legislation had not obstructed the victory". An irony which Luigi Messedaglia noted when he claimed that pellagra "has not been cured; it has cured itself spontaneously … while scientists were still arguing about its causes … making fun of "further deeper studies" and of "doctrinal alliances"" (Messedaglia 1927, p. 159). Not the first time in medical history that successful action had been taken for the "wrong" theoretical reasons.

Malaria and Quinine: The First Magic Bullet

Malaria is an ancient scourge of Italy, well known since Roman times (Sallares 2002). It was one of the most important public health problems for Liberal Italy and continued for a full century after unification, although the disease had disappeared from the rest of Western Europe and was essentially a tropical disease and a colonial problem. Malaria, along with pellagra, was a typical peasants' condition "closely associated with the social question" (Celli 1903, p. 3) and, according to Francesco S. Nitti, "the single factor accounting for the backwardness of the Italian South" (cited by Corti 1984, p. 646).

In unified Italy, malaria captured national attention in 1879 with the Parliamentary Railway Commission's report by Luigi Torelli, who drafted in 1882 the first map of malarial Italy (Torelli 1882). At the end of the 1880s estimates were that about two million people were infected or re-infected by malaria, and more than 20,000 died each year. Malaria was spread throughout Italy (only 2 out of the 69 provinces were officially free of malaria), but in terms of numbers and severity, there were two malarial Italies. In the north, the prevalence was lower and the malarial forms milder, while in the south about one-third of the cases were of the most severe falciparum form. The two Italies of malaria had also different ecological models. In the north, the man-made rice fields of the Po Valley intensively worked by casual and unskilled day labourers, predominantly women (Snowden 2006). In the south, the natural swamps created by rivers over-floating their banks, drainage, and irrigation ditches, coastal stagnant ponds of the large and depopulated estates of the Roman campaign, Apulia, Sicily, and Sardinia (Corti 1984).

After a few major attempts at *bonifica* (land reclamation), Giolitti's governments adopted a strategy of malaria eradication based on mass preventive quininization, a vertical programme which gathered direction from the discoveries of the Roman school of malariology on the complex relations of parasites, mosquitoes, and people which was legislated by a series of state quinine laws from 1901 to 1907. Malariology was for a long time the glory of Italian medical science, and the Roman school, with Ettore Marchiafava, Giovanni Battista Grassi, and Angelo Celli, dominated malaria studies at the turn of the century. That was an age ripe with scientific nationalism, and the Roman School was frequently celebrated as the Italian response to Louis Pasteur for France and Robert Koch for Germany. The School was divided by scientific, personal, and political controversies both in Italy and with the international community (Corbellini 2006). One of the bitter controversies in the history of medicine pitted Giovanni Battista Grassi, a zoologist from the University of Rome against Ronald Ross, a medical officer of the British Colonial Medical Service (see for example Grassi 1901). The dispute was about the priority in the discovery of the mechanism of transmission of malaria and was compounded by the desire for academic pre-eminence, differences in methods of research (Grassi took pride in applying the systematic and comparative method of zoology), and also nationalistic undertones. Eventually Ross, with the arbitration of Koch, received the 1902 Nobel Prize in medicine. The sharp exchange however continued and extended to the worldwide medical Academy, including Patrick Manson, Camillo Golgi, Luigi Pagliani, and many others creating "the network of intrigue" depicted by Lewis Hackett years later (Stapleton 2000). Ronald Ross continued to condemn vehemently "the Italian pirates", entitling a chapter of his memories "Roman brigandage", where he also described a situation of all against all (Ross 1923, p. 396).

As with pellagra, malaria campaigns were often intermingled with the rise of peasants' mass movements, both in the rice fields of the Po Valley and in the great estates of the *Campagna Romana* and Foggia in Apulia. These were the strongholds of the campaign against malaria and of peasants' mass movements (Snowden 2003). The Socialist Party and the National Farm Workers' Federation (*Federterra*) took advantage of the malaria campaign as a major aspect of their political initiative. *L'Avanti*, a socialist daily newspaper, claimed that "socialism and quinine" advanced together among the peasants, the traditional outcasts of governmental social policies. This often gave strong political overtones to scientific controversies as when *Federterra* officially endorsed Celli's prophylactic use of quinine as an example of "socialist therapeutics" (*L'Avanti!* October 12, 1909). Women contributed to political mobilization and substantially influenced the malaria campaign. In the northern rice fields, about 80 per cent of the seasonal workforce of low-skilled weeders were women (Gentili Zappi 1991). In the *Campagna Romana*, Anna Celli-Fraentzel and her *Unione Femminile* organized the first peasant school in 1904, and teachers soon joined *medici condotti* in the movement for redeeming the *Campagna* and its peasants (Snowden 2003). The malaria law of 1907 contained important dispositions protecting women and children, banning night work and limiting the workday to nine hours, prohibited work in the last month of pregnancy, and mandated two breaks to breastfeed children.

The quinine programme was administered by the *Commissione di vigilanza sul Chinino di Stato*, which reported to the Ministry of Finance and was masterminded by Angelo Celli, along with the Society for the Study of Malaria founded in 1898 and chaired by Giustino Fortunato. The first legislation concerning malaria was based on two bills proposed by Leo Wollemborg and Sidney Sonnino in 1900 aimed at making quinine readily available in malarial areas at a very low price by establishing a state monopoly over its production by the *Istituto Farmaceutico Militare* in Turin. Distribution was through stores selling monopoly goods, such as tobacco and salt, and funding came from a tax levied on employers and landowners. Across time, legislation progressively expanded its reach in terms of coverage and instruments of intervention. New laws established the right of those living or working in designated malarial districts to receive state quinine free of charge and enabled municipalities and charities to buy quinine tablets at very low prices which could also be distributed to the poor, now covered by Giolitti' s health law. This was similar to a sort of English "poor rate", which Celli described as "a piece of socialism coming true" (Celli 1903, p. 3). Moreover, government and municipalities, railway companies, and their contractors for public works in malarial districts were required to protect their employers' dwellings with mechanical barriers, and malarial deaths in public work labourers not treated with quinine were taken as occupational injuries. Tablets were progressively distributed in rural health stations and dispensaries, which often offered also educational programmes for children and adults.

The prophylactic use of quinine, which was given not only to the sick but also to healthy, non-malarial subjects, was the essential tenet of Celli's strategy for attacking malaria. It was also the main basis of division within Italian malariology. Italian malariologists agreed for the most part on the science of malaria, but were divided over anti-malarial strategy, and each school strenuously defended its own approach (Fantini 1998). Grassi insisted on mechanical house protection with wired gauze, supported a "radical cure" through "blood sterilization" of infected persons in the inter-epidemic season, but was against Celli's strategy of prophylactically using quinine for healthy persons, supporting instead Koch's method of "blood sterilization" in infected patients. Celli introduced and tirelessly defended the practice of mass prophylactic quininization, although he also assumed that "malaria can only be won solely by attacking it from every angle. *Unum facere et alterum non omittere* (do one thing but not dismiss the others) has been my battle cry for a long time" (Celli 1906, p. 707). In fact, Celli had experimented on preventing mosquito bites by mechanical screening and the use of repellents, as well as on treating a small number of persons with quinine (Celli 1900). On these grounds, he argued for broad social, environmental, and behavioural interventions in contrast to the reductionist Grassi law, which limited the interpretation of malaria infection to the equation "mosquito plus infected persons" (Celli 1906). The constitution in 1909 of the *Lega Nazionale contro la Malaria* (National League against Malaria) chaired by Camillo Golgi institutionalized the fierce antagonism with Celli's and Fortunato's *Società Italiana di Studi della Malaria* (G.C 1909). With the support of Guido Baccelli, Giambattista Grassi, Nicola Badaloni, and the tacit backing of the *Direzione generale di sanità*, the *Lega* dismissed the strategy of quinine prophylaxis as "scientifically bankrupt" (Tropeano 1908, p. 144) and advocated broad social and agricultural reforms as the only proper solution against malaria. The division focused on the prophylactic or therapeutic use of quinine and started a period of official scrutiny of the impact of quinine prophylaxis in various parts of Italy by a series of Commissions set up by the *Direzione generale di Sanità*, which determined its decline.

The decline in mortality for malaria from 710 deaths per million population in 1887 to 490 in 1900 and 57 per million in 1914 was spectacular. The trend of malaria deaths between 1900 and 1914 showed a remarkable and steady decline, interrupted only by occasional recrudescences (Fig. 2.2). The decline was associated with the expansion in the distribution of state quinine, which had been made available free to an increasing number of sufferers or exposed to infections. This had also produced nice profits, which were reinvested in the programme (Celli 1906). The achievements of the state quinine campaign had engendered great optimism and encouraged seeing quinine the first "magic bullet" against one of the greatest Italian scourges. However, as we have seen, experts were divided over the interpretation of the role of quinine in malaria control. The steep increase in the number of cases and of deaths from malaria with the outbreak of the First World War revealed that the goal of eradication

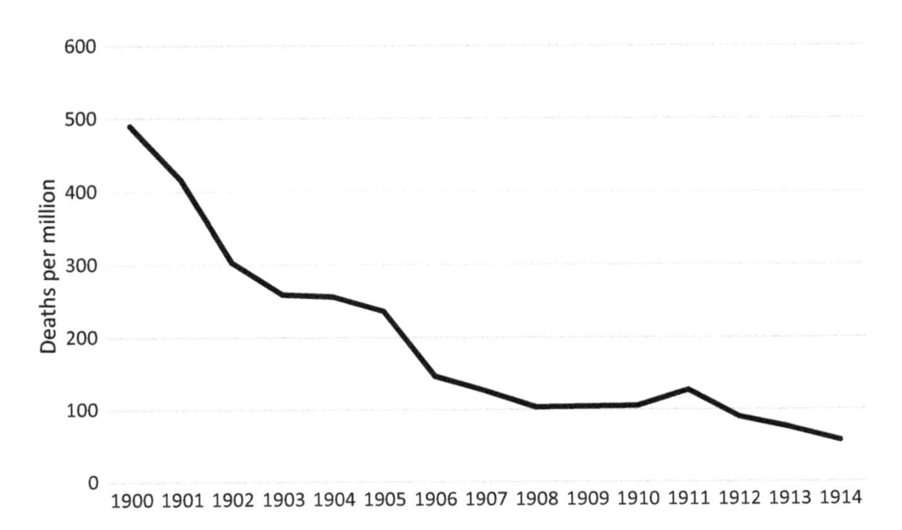

Fig. 2.2 Mortality from malaria, 1900–1914, Number of deaths per million population. (*Source*: ISTAT, 2011)

of the disease by quinine was a mirage. At the end of the Great War the number of deaths which had decreased from 600 per million population to less than 50, increased again to 320 per million in 1919, when Italy still had some 2 million cases per year (see page 76). Quinine had reduced the number of deaths but had only had a limited impact on morbidity (Ministero dell'Interno 1924). The fascist regime would engage in its "battle against malaria" relying on quite a different strategy, based on *bonifica integrale* (integral land reclamation), where quininization had only a limited role (see page 95).

Doctors, Medical Care, and Work Safety

Medical services in the Liberal state were essentially local in nature and depended entirely on local resources. Private and public provision coexisted for both domiciliary and residential care and relief. In Italy in 1877, there were 18,044 "approved" doctors, or 6.1 doctors per 10,000 population (Raseri 1878). Relative to the population, the Italian rate was twice the number of doctors as in France (2.9, excluding sanitary officers) and Germany (3.2), and close to that observed in England (6.0 per 10,000 population). Doctors were concentrated in the largest communes, with 1 doctor for every 880 residents, compared to 1 doctor for every 1780 residents in rural areas. There were also a small number of phlebotomists (1 every 10,000 residents), based mostly in the rural areas of the southern regions. About half of the approved doctors practised privately, and these were mostly concentrated in the cities.

"Public" provision drew on an intricate network of agencies including the Catholic Church, municipalities, private charities and individual philanthropists, and, in a very limited extent and mostly in the bigger cities, self-help and mutual aid through *Società di Mutuo Soccorso*. This envisaged a mixed economy of healthcare in which religious and private, philanthropic charity usually overshadowed the relief provided by public institutions, which essentially acted as providers of last resort. The central state had a limited role, both organizationally and financially, and fragile relations existed between central and local governments. This left wide scope for local initiative and, usually, inertia. Prefects often complained in their reports about the apathy of municipal administrations in undertaking the policies legislated, but not funded, by the central government (Patriarca 1994).

Healthcare organization revolved around *condotte mediche*, managed and financed by communes, which were responsible for the health of their residents. Municipalities appointed and paid *medici condotti*, either alone or in combination with other communes (*consorzi*). *Medici condotti* oversaw sanitary measures and vaccinations and provided home medical assistance to the residents of the commune. Domiciliary care was kept separate from, and generally valued more than, institutional care provided by *Opere pie ospitaliere*. The landmark survey on the sanitary conditions of the Kingdom's communes provides a detailed description of healthcare organization in the early 1880s, before the enactment of Crispi's laws (MAIC 1886). Out of the 8259 communes, 7564 held a *condotta*, in 108 domiciliary relief was provided by Opere pie, and 587 had no medical service whatsoever. Communes without *condotte* were particularly concentrated in Piedmont and Sardinia, both former parts of the Piedmontese Kingdom of Savoy, validating the strong words of Gaetano Strambio against Piedmont organization and its health policy (see page 9). About two-thirds of the *condotte* (4154) covered the whole population of the commune (*condotte piene*), while the others provided only for the poor. Municipalities' obligatory expenses for condotte mediche (including home birth attendance by about 9000 *levatrici*—midwives) amounted to 11.4 million lire, compared to 18.4 million for clearing nuisances and lighting, and 9.8 million for land and urban wardens. *Condotte* were run by 8585 doctors (or just one per condotta on average), while there were 8983 private practitioners. Private practitioners were concentrated in the major cities: Naples, Genoa, Turin, and Milan accounted for over 13 per cent of private practitioners, with 3 per cent of the national population (Frascani 1986, p. 635). By 1905, private practitioners had increased by 39 per cent and *condotti* by only 20 per cent. About 12 per cent of all doctors were receiving a salary or, more frequently, steady payments from hospitals, in the process of their medicalization (Frascani 1982). Five years later, the Chamber debated "the period of crisis of the *condotta medica*" (Celli Chamber May 31, 1910), noting that the number of doctors had increased, but parts of Italy were still poorly served and the situation was worsening. In Sicily, for example, there was one doctor for every 9000 inhabitants, compared to one for every 8000 five years earlier (Sanarelli Chamber May 31, 1910). The problem was found again in the politics of the communes that evaded the legislation on obligatory expenses, and proposals were advanced to transfer the service to *provincie*. This was also considered by the fascist regime, after the First World War, but promptly dismissed.

Medici condotti had not gained the status and reputation they expected since unification (Frascani 1982a). In 1905, a report from Britain defined "the incomes of medical men … exceedingly low, except for those practicing in the big and wealthy northern cities, and the consulting physicians, particularly surgeons in hospital who charge fees for their operations" (Burton-Brown 1905, p. 1209). Their social status was also generally poor. Noting that graduates in medicine were usually not "persons of birth", Burton-Brown observed that "it is only those who achieve exceptional renown who are admitted to a social status with which the learned professions are associated in England" (Burton-Brown 1905, p. 1210). The medical profession as such did not enjoy much political power either. A few doctors were elected in the chamber and a fair number were appointed to the senate (Montaldo 2011), but few medical men (there were no women in Parliament) occupied prominent political positions, except Guido Baccelli, who was three times minister. Except in part in the Giolittian era, they were not very influential in passing health legislation to achieve their goals. At unification, the *Associazione Medica Italiana* sought to codify scattered pieces of legislation to provide a coherent body of sanitary standards that would guarantee uniform health organization across the country. However, it took twenty-two years for a public health bill to become law. Because of its divisions, the medical profession also failed in determining which issues entered the political agenda, as with the case of Pagliani's School of Public Health. Moreover, at the national, as well as the local level, they never succeeded in transferring power on health issues from political authorities, such as the mayors and municipal councils, to professional bodies, which continued to maintain only advisory functions.

Condotte were the rather exclusive source of a regular income for doctors and the place where they fought for occupational stability, professional autonomy, and political influence. This uniquely influenced the development of medical associationism. The fight against quacks was not a priority in Italy in the second part of the nineteenth century (Malatesta 2011, p. 65). Phlebotomists were a small minority, and the public health law of 1888, which granted monopoly to graduates from the faculties of medicine, abolished their profession. Moreover, social insurance was introduced only late in Italy, and doctors did not organize against the mutual-benefit funds as in Germany. Italian medical trade unionism had local authorities as its counterpart. *Medici condotti* were the first to organize in 1874 into a national organization, the *Associazione Nazionale dei Medici Comunali*. Initially, the aim was to provide mutual assistance to doctors in their negotiations and frequent litigations with local authorities and to lobby the central government for legislating their occupational stability and professional autonomy. This was substantially achieved first with Giolitti's amendments to Crispi's law in 1904, which provided relative stability, and then by the abolition of condotte piene (see page 22).

The professional policy changed with the *Associazione Nazionale Medici Condotti* (ANMC) founded in 1902 (Detti 1979). This new organization was essentially a vehicle of propaganda for social reform, closely related to the socialist party. Particularly in the rural areas, *medici condotti* practised a sort of civil ministry, along with elementary school teachers, the equivalent of lay priests equipped with scientific training. They preached an "evangelical socialism" combining humanitarianism and scientism as a sure way to secure progress (Levy 2001, p. 196; Detti 1979, p. 3). Some just saw in what they called socialism "the natural convergence between social laws and the laws of hygiene" (Sormani 1893). *Medici condotti*, as we have seen, participated in the rural movements of landless labourers against malaria and pellagra in the Po Valley, the Pontine Marshes, and Apulia, and rapidly gained a reputation of "sovversivism" (Cogliolo 1899). They were less directly involved with specific problems of industrial diseases and work safety, which they tended to ground in broader social problems. The factory was primarily seen as a focus for a much broader spectrum of urban problems, where urban and industrial causes of ill health intertwined (Dodi 1978). On the other hand, social diseases of the rural areas such as malaria and pellagra were defined, under certain conditions, occupational diseases. Both issues were incorporated into a very broad concept of public health, which lacked specificity and was easy prey of specialisms from engineers, architects, and urbanists (Giovannini 1996).

This explains, in part, why health and safety in the workplace was an underdeveloped issue in Italy, despite its long tradition in occupational health. The tradition dates to the late 1600s with Bernardino Ramazzini, who examined the diseases in the main occupations of his time (Carnevale and Baldasseroni 2000). The study of the diseases of specific crafts ("artificum" in Ramazzini's own word) was long maintained, also because of Italy's belated industrialization (see, e.g., Mantegazza 1881). Conditions of work, work safety, and industrial accidents were generally treated as part of the "social question". Proposals for including workers' legislation in the Codice Lanza came not from medical hygienists but from conservative social reformers, such as Luigi Luzzatti, an early leader of German state socialism in Italy. Luzzatti's proposal was inspired by the British Factory

Acts of 1833 and 1844 (Bartrip and Fenn 1983) and was confined principally to regulating child labour in the textile industry, the prevalent industrial sector of the time. Industrialists successfully opposed this legislation, claiming its negative effects on international competitiveness of the Italian industry and its disruptive impact on industrial relations. A flavour of the arguments comes from the exchange between Luzzatti and Alessandro Rossi, senator and the biggest and most active textile industrialist, "a paternalist with an avowed horror of emerging socialism, arguing for Italian modernism based upon English and American models" (Rapp 1972, p. 946; see also Lanaro 1971). According to Rossi, the British law, and by implication its Italian counterpart, "broke whatever relation of humanity and love links the owner to the worker and put both under an Inspector paid by the Government" (Luzzatti 1877, p. 330). From then on, issues of safety in the workplace were incorporated into the social question and ignored by health legislation. Crispi's public health law, for example, dealt only with the environmental effects of factories, which were then mostly located within city limits and treated as matters of traditional concern by hygienists (Frascani 1982b).

The first significant piece of work legislation was passed only in 1898 and concerned workers' compensation for industrial accidents. The law made compulsory insurance against recognized work-related risks the basis for workers' compensation, instead of the demonstration in court of employer's negligence. The law allowed for employers' free choice of insurance in a competitive market of private insurance companies, including those set up by associations of the employers, which made it acceptable to industrialists (Romano 1982). The law was a progress in principle but was weakly enforced and severely limited in scope. As usual with labour legislation, coverage was fragmentary and agricultural workers were excluded, and remained so until the law of August 1917, which was implemented only in 1919. After the weak legislation of 1886, specific factory legislation regulating hours and conditions of work for children and mothers dates from the early 1900s. The 1902 law forbade children under the age of nine from working in industry (agriculture was again excluded) and those under the age of ten to work on mining. This was also intended to help children attending elementary school, one of Luzzatti's aims twenty years earlier. A few years later a Factory Inspectorate was established to supervise the implementation of the law on industrial accidents and on certain minimum standards of industrial safety. Initially, the inspectorate was composed only of engineers, and medical doctors were excluded. This best describes the marginal contribution and disposition of doctors in general, and of the hygienists in particular, on the issues of workers' health and safety. The role of the medical profession was the focus of the National Commission established by Guido Baccelli as Minister of Agriculture, Industry, and Commerce "to implement the means for preventing diseases and disabilities caused by the practices of arts and crafts", which was composed of doctors and hygienists, including Angelo Celli and Luigi Devoto, but resulted in nothing (Baldasseroni and Carnevale 1999, p. 433). Medical commissioners considered in the main the contribution of the workplace and of labour more generally to common diseases, instead of focusing on conditions caused by specific industrial risks. Solutions proposed were divided between a specific insurance for occupational diseases, just including them into existing accident insurance, and a general sickness insurance. The medical interest concerning occupational diseases rose with the First International Congress on Occupational Diseases, held in Milan in 1906, as part of the great International Exhibition celebrating the Italian industrial take-off (Carnevale et al. 2006) and the subsequent first Italian Congress. The high moment of the International Congress was Luigi Devoto's announcement of the decision of the municipality to open in a few years a *Clinica del Lavoro* (Clinic of Work) in Milan. The name was chosen to stress its interest in both the causes of disease in work and the effects of diseases in workers.

The opening in 1910 of the *Clinica del Lavoro* in Milan, the main centre of industrialization in Italy, made occupational medicine a distinct branch of academic medicine (Nenci 2008). *Clinica del Lavoro* was part of a broader project to make Milan the centre of a network of scientific institutes taking care of the postgraduate education of Italian doctors, "the missing link of medical studies" in Italy (Zocchi 2008). Luigi Devoto, its charismatic leader, shared the generic humanitarian scientism of the time and envisaged a scientifically based work legislation, where "a code, emanation of the laws of biology and physiology" would nurture a corporative social pact "between the government, industrialists and workers" (Devoto 1929). The *Clinica del Lavoro* has been described as marking "a meeting point between the more enlightened middle classes and the more reformist wing of the working-class movement" (Berlinguer and Biocca 1987, p. 455). The Clinica was financed by philanthropists of the Humanitarian Society, the Commune of Milan, local industrialists, and banks. It eventually obtained the benevolent neutrality of the central government in its feud with the University of Pavia, one of the main centres of academic medicine. Local trade unions supported the initiative, but a long and heated debate split the socialist party both at the local and national levels (Carnevale 2002; Calloni 1988).

The Rise of the Italian Hospital: Change Without Reform

Opere pie ospitaliere were local, self-governing institutions at the intersection of religious charity, municipal authority, and, eventually, scientific medicine. Their developmental path from charities for the relief of the indigent, widely perceived as "gateways to death" (Rosenberg 1987), into medical institutions embodying "modern medical science" (Vogel 1980, p. 77) was mostly dictated by local circumstances and cannot be easily traced back to a common pattern. Hospitals and other health-related organizations were the richest and most powerful of the *Opere pie*. At the time of unification, *Opere pie ospitaliere* were only 5.3 per cent of the over 17,000 *Opere pie*, but their patrimony of over 380 million lire accounted for more than one-third of their total assets. According to Enrico Raseri their 1878 survey accounted for 1435 organizations providing residential medical care of which 1138 were *ospedali per infermi* (hospitals for sick patients), 62 *ospedali per cronici* (hospitals for chronic patients), 18 *case di maternità* (maternity houses), 15 *manicomi* (asylums), and 102 *brefotrofi* (foundling homes) (Raseri 1883). In addition, he mentioned an unknown number of *case di salute* (health homes), which were privately owned, and a few hospitals owned by municipalities and provinces. In 1880, the survey of the Royal Commission reported that out of 21,866 charities, 1187 were *Ospitali* with a patrimony of over 600 million lire. They included 245 *ospizi di mendicità* (shelters for beggars), 100 maternity and children's accommodations, 57 institutions for "chronic" and "incurables", 28 for the deaf and the blind, and 14 asylums for lunatics (Silei 2003, pp. 109–115). Furthermore, according to the survey on the *Condizioni sanitarie dei comuni del Regno* of the early 1880s, out of 1176 hospitals (numbers rarely matched from one survey to another), 88 per cent were concentrated in the central and northern parts of the country. Moreover, 133 hospitals in the ten largest cities accounted for 54 per cent of 57,765 total beds, averaging 239 beds each (332 in the north), against an average of 25 beds for the rest of the 1043 hospitals (MAIC 1886).

Hospitals were a foundation of the local economy. Returns on the patrimony and donations provided the large majority of hospital income, along with municipal grants and transfers for services provided to their poor residents. While in Bologna, Florence, Rome, and Naples hospitals drew their income in the main from real estate and public debt securities, those in Lombardy and Piedmont thrived on rents from extensive and very productive dry farms and the wetlands of the rice-belt. Senator and Count Stefano Jacini acutely observed in his *Relazione generale sulla Inchiesta agraria* that "some institutions of beneficence founded with the specific goal of providing for the sick with their tenancy contracts contribute to bring the misery causing so many diseases … they seem to say: let them get sick now and we will provide for them later" (Jacini 1885, p. 78).

Boards of administrators governed hospitals subject to conditions inscribed in their *tavole di fondazione* (foundation tables or charters), which determined who should benefit and in what form, appointed doctors and supervised the management of day-to-day operations. Attempts of various governments to exert control over the performance of *Opere pie* and increase their efficiency had not helped in introducing modern methods of uniform cost accounting as described in Britain as early as the 1850s (Sturdy and Cooter 1998; Robson 2008). Sandra Cavallo's studies on "the motivations of benefactors" as well as administrators show that hospitals remained useful means of patronage and that participation in their management favoured the creation of networks of interests (Cavallo 1989). The honorary medical staff had an obvious interest in strictly limiting admissions to the sick poor to ensure that charitable medical provision was not an alternative to paying for their services in the open marketplace.

None of the major sanitary laws of the liberal state and only a few of their *Regolamenti* dealt with hospital organization and their management. In 1874 the *Regolamento* of the law of 1865 prescribed the isolation of patients with contagious diseases, provided an indicative standard of fifty beds per doctor, and recommended that hospitals adopt their own *regolamento* (charter), which was easy prey for administrators and benefactors to the detriment of the doctors' role in the hospital (Verardini 1873). The French doctor August-Gabriel Millot provided a vivid description of their unsanitary conditions in the 1870s: "Italian hospitals are generally poorly maintained and badly built and almost all are inadequate for their patients' needs. Luxury reigns, but only in the façade" (Millot 1875, p. 112). Inside "air is scanty, cesspools exhale stinking smells, and ventilation systems, similar to those in use in Belgium or England, are wanting" (Millot p. 115). Hospital Charters show that straw mattresses were the norm, and sheets, when available, were seldom changed during the stay of the inmates, who were allowed to bath about once a month. Italian hospitals, and *Opere pie* in general, did not officially follow the English poor law principle of less eligibility, but the inmates were usually required to perform domestic services and attend other patients as "nurse inmates" or "assistants" to promote their own morality and prevent "pauperization", as well as in the interest of the hospital economy. Giuseppe Ruggi, surgeon in the Ospedale Maggiore di Bologna in the late 1870s, reported the impact of poor hygienic conditions on patients' health: "piemia, septicemia, gangrene and erysipelas polluted the operation room, and from time to time, sick patients died one after another, with no treatment available" (Ruggi 1924, p. 65).

Evidence of the general veracity of these observations can be found in the complex story of the Ospedale Maggiore in Milan, one of the richest and biggest hospitals in Italy at the time. The *Commissione provinciale d'inchiesta* (Commission of Enquiry) to oversee the conditions of the municipal hospital was chaired by Gaetano Strambio and included, among others, Gaetano Pini and Malachia De Cristoforis, pioneers in hospital design and organization. It reported that behind the hospital's magnificent façade one found crowded wards, poor light, and ventilation, lack of places for isolating contagious patients and for performing antiseptic surgery (*Relazione* 1882; see also Forti Messina 2003, p. 98). The conclusions stressed the stark contrast between expenditures on monumental buildings designed to evoke the social recognition of benefactors' munificence with the necessities of modern assistance. "The hospital is going out-of-date… it has remained the same when everything changed, when the breath of science has combined with the push from philanthropy to open new paths to public assistance for the sick poor" (*Relazione* 1882, p. 50). A crucial issue was the need to build new hospitals based on modern architectural and functional criteria instead of straining to renovate the old ones, which "once a glory of the ancient beneficence, they are now an offence to hygiene and a threat for the sick".

Progress in nineteenth-century medical science transformed medical practice in various sectors, from hygiene to clinical care. Developments in anaesthesia and antisepsis raised surgery as a crucial aspect of hospital practice, expanding its curative capacities and contributing to the decline of "hospitalism" (i.e., what we now call hospital or nosocomial infections) and hospital mortality (Lister 1867). However, cultural resistance, political issues, and financial problems compounded and confused the diffusion of "listerism" in its various forms contributing to the "implementation gap" of scientific progress in Italy as well as in Britain (Pennington 1995). In Bologna, Ruggi complained about "the very high cost of medications and the need for frequent dressings' renovation" as "the most inconvenient fact" of adopting listerism in surgery (Ruggi 1924). Complaints in Bologna were echoed in Milan, where "doctors' support of this system was opposed by the administration because of its cost" (Corradi 1882, p. 86). When the Commission's Report on the Ospedale Maggiore of Milan was submitted to a number of progressive national leaders of the hygienic movement, from Agostino Bertani and Luigi Pagliani to Arnaldo Cantani from Naples and Giacinto Pacchiotti from Turin, commentaries stressed the need for radical renovation. Spatial distribution and functional organization of the modern "acute" hospital ought to be shaped by the principles of asepsis and the necessities of surgery and of isolating patients with contagious diseases, with obvious implications in terms of architectural design, internal organization, and doctors' power in its administration.

In the early 1880s, an intense debate developed concerning hospitals' social function and their governance, a crucial step towards "hospitalizing medicine and medicalizing the hospital", to borrow Rosenberg's terms (Rosenberg 1979). A defining event in the separation of poor relief from medical assistance and the differentiation of hospital from domiciliary care was the International Conference on Public Health and Beneficence held in Milan in 1880, where Alfonso Corradi compared domiciliary and hospital care and theorized a separation of acute hospital care from long term and elderly care (Corradi 1882). A few months later, the Genoa IV Congress of the Italian Medical Association (AMI) took instead a more "inward vision" (Rosenberg 1987) and focused on "*il miglior governo degli ospedali*" (optimal hospital governance), summarizing a decade-long debate on the role of the doctors in hospital administration initiated with the Bologna charter of hospital organization (Zucchi 1875).

The rise of hospital care and increased hospital admissions had fuelled an international debate on whether the hospital or the patient's home was the optimal place for medical care. Delegates to the Milan international conference came primarily from France, which was considering a shift from church-based charité to state-directed biénfaisance, structured into Local Bureaux for domiciliary care and hospitals and hospices for acute and long-term residential care respectively (Domin 2019). The attention of the lay and the catholic press was captured by the politically charged Church-State conflict implicit in the terms of reference of the recently appointed Royal Commission on the transformation of *Opere pie* chaired by Cesare Correnti, who also participated in the conference (see page 16; a discussion of the arguments against the secularization of Opere pie menaced by the Commission is in Quine 2002, p. 52). This partially eclipsed the discussion of the main paper delivered by Alfonso Corradi, a long-time leader of Italian doctors, President of the Medical Faculty, Rector of the University of Pavia, and President of Italian Society of Hygiene. Corradi's presentation and the ensuing discussion were a bold and shameless claim for the superior value of domiciliary care over hospital admission in clinical, economic, and moral grounds (Corradi 1882). Based on the evidences from a companion paper including over 13,000 patients followed by 13 *medici condotti* (Corradi 1881), he argued that domiciliary care consumed fewer resources, cost less, allowed doctors to combine preventive and curative medicine for the patient and their family, and prevented "charity abuse" typical of hospitalization, which instilled a pauperizing mentality in the recipients of unearned benefits. He recognized, however, the necessity of admission to hospital for isolating contagious patients, when home was inappropriate or in case of surgery. Since the hospital's primary mission was to treat acute cases, his imperative was to "separate" acute and treatable cases from chronic and incurable conditions. Separation applied also both between and within hospitals. Between hospitals, separation nurtured the concept of a hierarchical system differentiating hospitals by function, with small rural facilities devoted to simpler cases and relief of the chronic and incurable, and more complex and surgical cases concentrated in major urban hospitals. Within major hospitals, the growth of specialism as a vertical division of labour partitioned the practice of hospital medicine by disease and technical procedures, induced doctors' specialization, and promoted the rise of skilled allied personnel, such as nurses and other technicians.

One of the main conclusions of the Congress was on the need to develop appropriate information on the activity of hospital care and its outcomes. This prompted Enrico Raseri to organize a clinical information system to provide *statistiche ragionate* (reasoned statistics) of the end results of "certain disease conditions under different treatments or of the same treatment in different conditions" (Raseri 1883, p. 127). As the Congress had claimed, these statistics would prove that "the reasons of charity converge with the true progress of science" and "*sanitas* combines with *caritas*". Hospitals were required to provide the central statistical office with standardized tables, extracted from their own Registries, detailing age, gender, days of stay, disease(s) treated, and patients' status at discharge (cured, improved, steady, worsened, or dead). Diseases were coded according to a specific list of 169 hospital conditions, different from those used for tabulating causes of death. The project attests to a crucial passage in the purpose and the public image of the hospital, which is also reflected in the changing metaphors of its representation, from metaphors of charity and elevations of the souls of the beneficiaries and the beneficients to metaphors of science, outcomes, and efficiency. Raseri's grand scheme of "reasoned statistics" started on January 1, 1883, but floundered in a few years. Aggregate data on hospital admissions, their causes, discharge status, and days of hospital stay are available from 1883 to 1885, when the ambitious programme was terminated for "want of resources" (MAIC 1888). From then on, *Annali di Statistica* casually reported summary reports of ad hoc surveys based on questionnaires. Inconsistencies about the number of hospital facilities were the norm, as the definition of the hospital was still uncertain, and private hospitals were usually unaddressed.

Corradi's outward perspective in defining the hospital's social function at the International Congress in Milan was complemented by the inward approach of the Genoa Congress of AMI, with the goal of strengthening the doctors' role in the internal hospital organization. In the 1874 *Congresso Medico di Bologna*, AMI had approved a sixteen-point charter for a modern hospital structure and organization. Among the main points, the charter envisaged appointing a *direttore sanitario* (medical director) to manage medical activities; recruiting *primari* (heads of clinic) through competitive examination by peers to halt appointments negotiated by the administration through informal networks of patronage; organizing hospital beds into (at least) five "divisions", run by "*medici speciali*" (special doctors) for lunatics, syphilis, and disease of the skin, eye, and female organs. Moreover, nuns should be replaced by skilled nurses trained at special schools annexed to major hospitals (Zucchi 1875). The model saw hospitals as "doctors' workshops", where resident physicians in possession of expert knowledge, assisted by professionally trained nurses, organized and provided specialized responses to specific disease entities, which had been selected as amenable to medical intervention according to Corradi's "separation" principle.

The debate focused less on the sixteen points, which were widely shared, than on the optimal strategy to secure its implementation through either national legislation or local initiatives and negotiations. Alternatives for hospital doctors were either the model of *medici condotti*, who had entrusted their expectations of status and income to state regulation or to continue the path of liberal self-governance and professional autonomy to follow the advancements of science, free from bureaucratic control. On the suggestion of Carlo Zucchi, medical superintendent of Ospedale Maggiore in Milan, the momentous conclusion of the Congress was for the latter, assuming that current hospital problems would fade away with progress and the relentless accumulation of medical knowledge and therapeutic innovation (Zucchi 1880 and 1881).

Both progress and legislation defied doctors' expectations. Doctors were coming to see hospitals not as sites for showing their benevolence and civic responsibility (a vision easily shared with lay administrators and benefactors) but as healing places for the application of scientific knowledge and medical expertise in treating specific diseases. Doctors' new vision conflicted with lay administrators and threatened not simply the structure of governance but the system of hospital charity itself. However, plans for medical dominance over their working conditions (the first and a clearly essential step of their process of professionalization) were countered by the increasing financial demands made upon hospitals' traditional sources of income, which strengthened the power of administration. Crispi's imperative of secularizing *Opere pie* and putting them under government control left hospitals submerged and confused amid a variety of very different institutions, and his 1890 law deferred to further regulation which never materialized. However, the process of secularizing charities and transferring their management to *Congregazioni di carità* had the indirect effect of municipalizing hospital governance and politicizing their administration. Moreover, the cost of hospitals' increasing medicalization prompted stricter relations with municipal councils and made hospital finances part of the local political economy.

A special survey on hospital beneficence performed between 1889 and 1892 provides information on hospital structure and organization at the time of Crispi's legislation (MAIC 1888). The survey identified 1158 hospitals, and a questionnaire was sent to the 889 with ten or more beds. Responders averaged sixty-one beds with an occupancy rate of just 63.5 per cent. Forty-four per cent also provided outpatient care and one-third had a pharmacy. Only ten hosted a nursing school. The medical staff was composed of 2871 doctors (averaging 3.2 per hospital), of which two-thirds were *primari*, supervising about twenty beds each and paid an honorarium averaging 1000 liras per year, higher than their assistants (757 liras), while *medici condotti* who were also engaged as hospital doctors earned 213 liras per year for their work in the hospital. An average of eight nurses per hospital (about half of them nuns) took care of four or five patients each.

Between the end of the century and the First World War, hospital admissions steadily increased, almost doubling from about 345,000 to more than 600,000 (Table 2.3). Hospitals increased more in size than in number, from 1176 in 1885 to 1376. The increase in size and admissions was higher and faster in the more rapidly developing urban hospitals of the northern regions, which further deepened the north-south divide. In the south, hospitalization rates only grew from 5 to 6 per 1000 population compared to increases from 18 to 22 and 19 to 25 per 1000 in the north and centre respectively (Frascani 1986, p. 165). Between 1885 and 1907, in major hospitals of the bigger cities hospital mortality fell by 18 per cent, from 12.0 to 9.8 per 1000 admissions, but increased by about the same amount (15 per cent) in rural hospitals (Frascani 1986, p. 148). The two opposing trends reflect both differences in case-mix, where rural hospitals admitted more chronic, older, and incurable patients, and to differential improvements in hospitals' hygienic conditions.

During the Giolittian age, hospital beneficence remained essentially an urban phenomenon concentrated in the big cities of the northern and central parts of the country. The growth of casual workers in the new labour market created by the development of big industry on the one hand and the diffusion of surgery on the other explain the massive increase in urban hospital admissions from the neighbouring *comuni*. This made

Table 2.3 Hospitals and their activity, 1885–1914, selected years

	1885	*1891*	*1907*	*1914*
Nr. of hospitals	1176	1158	1291	1376
Nr. of beds	57,765	n.a.	60,268	85,728
Discharges (×1000)	345	329	496	n.a.
ALOS[a]	26/37	n.a.	29	n.a.

Sources: Frascani (1986); Annali di Statistica (1895) and ISTAT Statistiche Ospedaliere (1934)

[a]Average length of stay

impractical the provision of *domicilio di soccorso* introduced by Crispi's law of 1890 and contributed to jeopardizing the financial stability of urban hospitals. A further source of expenditure was the increasing number and professionalization of the nursing and the technical staff, an essential resource for "modern hospitals" conceived as "organizations scientifically and operationally perfect" (Fiorilli 2015). The professionalization of the nursing staff intensified under the push of doctors' demand and the project pursued by some women' s associations such as the *Unione Femminile Nazionale* (National Women Union) of a modern occupation "particularly suited for women" (Celli-Fraentzel 1908). The growing demand for hospitalization, shrinking hospital income, and increased expenditures on drugs, technologies, and personnel made hospital deficits commonplace. Initially many major hospitals adopted contingent solutions proclaiming generalized *serrate* (lockouts) or accepting only patients from limited catchment areas defined according to the borders of their ancient origins. When hospitals realized that they were facing a structural problem and that charity could not financially cope with the cost of the new dynamism of hospital medical practice, they turned to the state for help. However, Giolitti stood with its traditional policy of fiscal conservatism: "the solution to the hospital question in Italy follows from the *legge comunale e provinciale* which makes it compulsory for *comuni* to take care of their sick poor… transferring to the State one of the most elementary duties of the comune is utterly inconceivable" (Giolitti Camera 15 marzo 1913). Diversification of hospitals' sources of income along the paths followed by American, Canadian, or British hospitals at about the same time encountered in Italy both professional resistance and structural problems.

American and Canadian hospitals supplemented their traditional income by opening to "patients of moderate means" and collected their fees for board and services as a significant new source to finance their modernization (Gagan and Gagan 2002; Stevens 1989). In Britain, the response to the lack of hospital coverage in Lloyd George's health insurance bill of 1911 was the growth of contributory schemes providing access to voluntary hospitals, the main sites of acute care (Cherry 1997). In Italy, financing the expansion of scientific medicine with revenue generated by paying patients was difficult to implement. Admitting middle-class patients clashed with the interests of *medici condotti* and particularly urban private practitioners. Both were threatened by the possible loss of paying patients who might be attracted by hospital services formerly restricted to poor patients. *Ordine dei medici* and *Associazione Nazionale Medici Condotti*

(ANMC) stood against "the hospitalization of the well-to-do". This was dubbed "a scourge for the medical profession invading every corner of Italy" and argued that "the hospital is for the poor and solely for the poor" while "for the well-to-do we have the *case della salute* (private clinics), whose diffusion hospitals must not hinder" (cited in Cherubini 1980, pp. 201–202). Major hospitals traditionally admitted, for small fees, patients "not poor enough to be eligible for *comuni* and *Opere pie* but not rich enough to afford a private physician at home or the admission to a private clinic" (for the *Ospedale Maggiore* in Milan see, e.g., Tanzi 1884, p. 62). In Naples, persons of good lineage (*nati ed educati civili*) "ashamed of mixing with the poor inmates" were admitted to *Ospedale degli Incurabili* in special wards for a fee covering only board and lodging. Later, most hospitals collected fees for ambulatory visits and outpatient clinics at about half or even one-third of those charged by private clinics that they shared with their doctors, inaugurating a practice that spread in the interwar years.

Sickness insurance was the solution officially pursued by the *Associazione Nazionale fra i medici degli Istituti ed Opere Pie Ospitaliere d'Italia* (National Association of Doctors of Institutes and Opere pie ospitaliere). Its president Agostino Carducci called for "a compulsory system of sickness insurance to ensure hospital access to everybody, not solely to the sick poor" as a new "inexhaustible and endless" source of hospital finance. His goal was shared by the First Congress of Hospital Medical Directors in 1913 arguing that "sickness general insurance" was "the unique potential relief of hospital budgets and the key for all other reforms we often asked for and proved necessary" (cited in Cherubini 1980, p. 205). This, however, was contrary to the priorities of national social policy that favoured old age and unemployment over sickness insurance and was dismissed as infeasible by both Giolitti and the socialists, who controlled ANMC. Chronic deficits pitted the boards of *Opere pie ospitaliere* against municipal councils and made hospital finance a part of the political economy of municipalities, upon which hospitals depended, as transfers from municipal coffers accounted for one-third to one-half of hospitals' annual budget. The civic nature of Italian hospitals and their close identification with local communities as catchment areas and political arenas grew stronger.

The First World War was the turning point in the history of Italian hospitals, after which they began to cater to a broader public and to focus on medical, as opposed to social, care. During the war, there was a rapid spread of major technical advances in medical and surgical care, particularly in orthopaedics and radiology, and the professionalization of nursing intensified, in collaboration between the Italian and the American Red Cross. These developments demanded increasing resources and left hospitals in a dire financial crisis as soon as wartime state financial support ended. By the end of the Great War, the old system of hospital assistance focused on the sick poor and based on beneficence, and municipal transfers had reached its limits. For a few months wartime policy seemed to be heading down the road towards sickness insurance, which however soon floundered again (see page 81).

References

Alfani, G. (2014) Le stime della mortalità per colera in Italia. Una nota comparativa. Popolazione e Storia, 2, pp. 77–85

Anonymous (1884) Il colera flagello e maestro. Civiltà Cattolica, 8, pp. 129–142

Anonymous (1885) La politica sanitaria dell'Italia nella epidemia coleriche, 1884–1885. Nuova Antologia, 84, pp. 639–666

Anonymous (1887) The hygienic renascence in Italy. BMJ, 2, pp. 728–729

Anonymous (1910) Italy. The cholera outbreak. Public Health Report, 25 (37), pp. 1286–1288

Anonymous (1911) The cholera situation. Public Health Report, 26 (39), p. 1459

Baldasseroni, A. Carnevale, F. (1999) Malati di lavoro. Storia della salute dei lavoratori. Bari-Roma, Laterza

Barbagallo, F. (2015) Napoli bella époque. Bari-Roma, Laterza

Bartrip, P.W. Fern, P.T. (1983) The evolution and regulatory style in the nineteenth century Britain Factory Inspectorate. Journal of Law & Society, 10, pp. 201–222

Berlinguer, G. Biocca, M. (1987) Recent developments in occupational health policy. International Journal of Health Services 17(3), pp. 455–474

Bosworth, R.J.B. (1979) Italy the least of the Great Powers. Italian foreign policy before the First World War. Cambridge, Cambridge University Press

Brice, C. (2002) "The King was pale". Italy's national-popular monarchy and the construction of disasters, 1882–1885. In: Dickie, J., Foot, J., Snowden, F. M. Eds, Disastro! Disasters in Italy since 1860. Culture, politics and society. New York, Palgrave, pp. 61–79

Brown, P.J., Whitaker, E.D. (1994) Health implications of modern agricultural transformations. Malaria and Pellagra in Italy. Human Organization, 53 (4), pp. 346–351

Burton-Brown, F.H. (1905) Italy. British Medical Journal, I, pp. 1208–1211

Calloni, M. (1988) I medici socialisti e la Clinica del Lavoro di Milano in una polemica del 1910. Sanità, Sicurezza e Storia, 1–2, pp. 333–351
Carnevale, F. (2002) Gaetano Pieraccini e la Clinica del Lavoro di Milano. Medicina & Storia, 2, pp. 109–118
Carnevale, F. Baldasseroni, A. (2000) The De Morbis Artificum Diatriba editions since 1700 and their legacy. Epidemiologia & Prevenzione 24(6), pp. 270–275
Carnevale, F., Baldasseroni, A., Guastella, V., Tomassini, L. (2006) Concerning the First International Congress on work-related illnesses. Milan, 9–14 June 1906. Success, News, Reports, Motions. Medicina del Lavoro 97(2), pp. 100–113
Cavallo, S. (1989) Charity, power and patronage in Eighteenth Century Italian hospitals. In: Granshaw L., Porter, R. Eds. The hospital in history. London, Routledge, pp. 93–122
Celli, A. (1896) Sconforti e speranze d'Igiene sociale. Discorso letto il 5 Novembre 1895 in occasione della solennne inaugurazione degli studi. Annuario della Università degli Studi di Roma. Roma,Tip.Flli Pallotta
Celli, A. (1900) The new prophylaxis against malaria: an account of experiments in Latium. Lancet, 156, pp. 1603–1606
Celli, A. (1903) La legislazione contro la malaria. Critica Sociale, 13 (4), pp. 3–14
Celli, A. (1906) Organizzazione della guerra alla malaria. Nuova Antologia, 121, pp. 707–717
Celli-Fraentzel, A. (1908) La donna infermiera in Italia. Roma, Tip. Bertero
Cherry, S. (1997) Before the National Health Service: financing the voluntary hospitals, 1900–1939. Economic History Review, 50 (2), pp. 305–326
Cherubini, A (1980) Medicina e Lotte sociali, 1900–1920. Roma, Il Pensiero Scientifico
Cogliolo, P. (1899) La questione dei medici condotti. Nuova Antologia, 83, pp. 508–516
Cohn, S.K. (2017) Cholera revolts: a class struggle we may not like. Social History, 42 (2), pp. 162–180
Corbellini, G. (2006) La lotta alla malaria in Italia. Conflitti scientifici e politica istituzionale. Medicina nei Secoli, 18, pp. 75–96
Corradi, A. (1876) Annali delle Epidemie occorse in Italia dalle prime memorie fino al 1850. Bologna, Tip. Gamberini e Parmeggiani
Corradi, A. (1881) Assistenza sanitaria dei poveri a domicilio. Annali Universali di Medicina, 253, pp. 118–143
Corradi, A. (1882) Relazione sul tema della beneficenza ospedaliera e sanitaria. Congresso Internazionale di Beneficenza tenuto in Milano nel 1880. Roma, Annali di statistica, Serie 2. vol. 14
Corti, P. (1984) Malaria e società contadina nel Mezzogiorno. In: Storia d'Italia. Annale VII, Malattia e Medicina. Dalla Peruta F. Ed. Torino, Einaudi, pp. 656–678

De Amicis, E. (1869) L'esercito italiano durante il colera del 1867. In: De Amicis, E., La vita militare: bozzetti. Firenze, Le Monnier, pp. 283–348
De Bernardi, A. (1984) Il mal della rosa. Denutrizione e pellagra nelle campagne italiane tra '800 e '900. Milano, F. Angeli
Del Panta, L. (1980) Le epidemie nella storia demografica italiana (secoli XIV–XIX). Torino, Lescher
Detti, T. (1979) Medicina, democrazia e socialismo in Italia fra '800 e '900. Movimento Operaio e Socialista, 2, pp. 3–50
Detti, T. (1982) La questione della tubercolosi nell'Italia giolittiana. Passato e Presente, 2, pp. 27–60
Devoto, L. (1929) La Clinica del Lavoro di Milano. Venti anni (1910–1929). Milano, Cordani
Dickie, J. (1999) Darkest Italy. The Nation and stereotypes of the Mezzogiorno, 1860–1900. London, Palgrave Macmillan
Dodi, L. (1978) I medici e la fabbrica. Prime linee di ricerca. Classe, 15, pp. 21–65
Domin J.P. (2019) Socialization of healthcare demand and development of the French health system (1890–1938). Business History, 61 (3), pp. 498–517
Evans, R.J. (1988) Epidemics and revolutions. Cholera in Nineteenth century Europe. Past & Present, 120, pp. 123–146
Fantini, B. (1998) Unum facere et alterum non omittere. Antimalarial strategies in Italy, 1880–1930. Parassitologia, 40, pp. 91–101
Fiorilli, O. (2015) "Un organismo scientificamente e praticamente perfetto". L'ospedale moderno e l'infermiera nel discorso medico del primo Novecento. Contemporanea, 18 (2), pp. 221–244
Forti Messina, A. (2003) Malachia De Cristoforis. Un medico democratico nell'Italia liberale. Milano, F. Angeli
Forti Messina, A.L. (1984) L'Italia dell'Ottocento di fronte al colera. In: Storia d'Italia. Annale VII. Malattia e Medicina. Dalla Peruta F. Ed. Torino, Einaudi, pp. 429–494
Fortunato, G. (1911) Le due Italie. La Voce, anno 3, n° 11, 16 Marzo
Frascani, P. (1982a) Il medico dell'Ottocento. Studi Storici, 23, pp. 617–637
Frascani, P. (1982b) La disciplina delle industrie insalubri nella legislazione sanitaria italiana (1865–1910). In: Betri, M.L., Gigli Marchetti A., Eds. Salute e classi lavoratrici in Italia dall'Unità al Fascismo. Milano, F. Angeli, pp. 713–735
Frascani, P. (1986) Ospedale e società in età liberale. Bologna, Il Mulino
Funk, C. (1912) The etiology of the deficiency diseases. Beri-beri, polyneuritis in birds, epidemic dropsy, scurvy, experimental scurvy in animals, infantile scurvy, ship beri-beri, pellagra. Journal of State Medicine, 20, pp. 341–368
G. C. (1909) Una lega nazionale contro la malaria. Nuova Antologia, 143, pp. 682–685

Gagan, D., Gagan, R. (2002) For patients of moderate means. A social history of the voluntary public general hospital in Canada, 1890–1950. Montreal, Mc Gill-Queen's University Press

Gandy, M. (1999) The Paris sewers and the rationalization of urban space. Transactions of the Institute of British Geographers, 24, pp. 23–44

Gentilcore, D. (2016) Louis Sambon and the clash of pellagra etiologies in Italy and the United States, 1905–1914. Journal History Medicine and Allied Sciences, 71 (1), pp. 19–42

Gentili Zappi, E. (1991) If eight hours seem too few. Mobilization of women workers in the Italian rice fields. State University of New York Press

Giovannini, C. (1996) Risanare la città: l'utopia igienista di fine Ottocento. Milano, F. Angeli

Grassi, G. (1901) Studi di uno zoologo sulla malaria. Roma, Tip. Accademia dei Lincei

Hamlin, C. (2009) "Cholera forcing" The myth of the Good Epidemic and the coming of the Good Water. American Journal of Public Health, 99 (11), pp. 1946–1954

Hardy, A. (1993) Cholera, quarantine and the English preventive system, 1850–1895. Medical History, 37 (3), pp. 250–269

Howard-Jones, N. (1975) The scientific background of the International Sanitary Conferences, 1851–1938. Geneva, World Health Organization

Huber, V. (2006) The unification of the globe by disease? The International Sanitary Conferences on cholera, 1851–1894. The Historical Journal, 49, pp. 453–476

Istituto Centrale di Statistica (1934) Statistica degli Ospedali e degli altri Istituti pubblici e privati di assistenza sanitaria ospitaliera nell'anno 1932. Roma, Poligrafico dello Stato

Jacini, S. (1885) Relazione finale sui risultati dell'Inchiesta. In Atti della Giunta per la Inchiesta Agraria e sulle condizioni della classe agricola. Vol. XV. Roma, Forzani & C. Tip. Senato

Kunitz, S.J. (1988) Hookworm and pellagra. Exemplary diseases in the New South. Journal Health Social Behavior, 29, pp. 139–148

Lanaro, S. (1971) Mercantilismo agrario e formazione del capitale nel pensiero di Alessandro Rossi. Quaderni Storici, 16, pp. 49–156

Levy, C. (2001) The people and the professors: socialism and the educated middle classes in Italy, 1870–1915. Journal of Modern Italian Studies, 6, pp. 195–208

Lister, J. (1867) Antiseptic principle in the practice of surgery. British Medical Journal, 2 (351), pp. 246–248

Livi-Bacci, M. (1986) Fertility, nutrition, and pellagra. Italy during the vital revolution. Journal of Interdisciplinary History, 16 (3), pp. 431–454

Logan, O. (2002) The clericals and disaster. Polemic and solidarism in liberal Italy. In: Dickie, J., Foot, J., Snowden, F. M., Eds. Disastro! Disasters in Italy since 1860. Culture, politics and society. New York, Palgrave, pp. 98–112.
Lombroso, C. (1869) Studi clinici ed esperimentali sulla natura, causa e terapia della pellagra. Bologna, Tip. Fava e Garagnani
Lombroso, C. (1892) Trattato profilattico e clinico della pellagra. Torino, F.ll. Bocca
Lussana, F. (1872) Sulle cause della pellagra. Milano, Tip. Richieldi
Luzzatti, L. (1877) Le Leggi sulle fabbriche in Inghilterra. Giornale degli Economisti 4, pp. 321–344
Maggiori-Perni, F. (1894) Palermo e le sue grandi epidemie dal secolo 16° al 19°. Saggio storico statistico. Palermo, Tip. Virzi
MAIC—Ministero dell'Agricoltura, Industria e Commercio. Direzione Generale della Statistica (1886). Le condizioni sanitarie dei comuni del Regno. Parte Generale. Annali di Statistica, Roma, Tip. Eredi Botta
MAIC—Ministero dell'Agricoltura, Industria e Commercio. Direzione Generale della Statistica (1888). Movimento degli infermi degli ospedali civili del Regno 1883–1885. Roma, Tip. Camera dei Deputati
Malatesta, M. (2011) Professional men, professional women. The European professions from the 19th century until today. London, Sage
Mantegazza, P. (1881) Almanacco Igienico popolare.Igiene del lavoro. Milano, Brigola
Melis, G. (1993) Amministrazioni speciali e Mezzogiorno nell'esperienza dello stato liberale. Studi Storici, 34, pp. 463–468
Messedaglia, L. (1927) Il mais e la vita rurale in Italia. Saggio di storia agraria. Piacenza, Federazione Italiana dei Consorzi Agrari
Millot, G. (1875) De l'Hygiene publique et de la chirurgie in Italie. Paris
Ministero dell'Interno.Direzione generale della sanità pubblica (1924). La malaria in Italia e i risultati della lotta antimalarica. Roma
Miraglia, N., Sormani, G. (1880) La diffusione della pellagra in Italia. In: MAIC Dir. Statistica, Annuali di Statistica, serie 2°, Vol. 15, pp. 207–210
Montaldo, S. (2011) Scienziati e potere politico. In: Scienza e cultura nell'Italia Unita. Annali 26. Cassata, F. Pogliano, C. Eds Torino, Einaudi, pp. 37–64
Mosso, A. (1884) Le precauzioni contro il colera e la quarantene. Nuova Antologia, 19, pp. 303–331
Munthe, A. (1909) Letters from a mourning city. London, John Murray
Nenci, E. (2008) Malati di lavoro. La Clinica delle malattie professionali di Luigi Devoto. In: Zocchi, P. Ed. La rete del perfezionamento medico. Milano scientifica, 1875–1924. Milano, Sironi, pp. 81–103
Niceforo, A. (1901) Italiani del Nord e Italiani del Sud. Torino, F.lli Bocca

Ogawa, M. (2000) Uneasy bedfellows. Science and politics in the refutation of Koch's bacterial theory of cholera. Bulletin of the History of Medicine, 74 (4), pp. 671–707

Patriarca, S. (1994) Statistical nation building and the consolidation of regions in Italy. Social Science History, 18, pp. 359–376

Pagliani, L. (1886) La polizia sanitaria in Italia di fronte alle epidemie di colera. Nuova Antologia, 15, pp. 464–486

Pagliani, L. (1894) Relazione intorno all'epidemia di colera in Italia nell'anno 1893. Roma, Tip. delle Mantellate

Pelling, M. (1978) Cholera, fever and English medicine, 1825–1865. Oxford, Oxford University Press

Pennington, T.H. (1995) Listerism, its decline and its persistence. The introduction of aseptic surgical techniques in three British teaching hospitals, 1890–1899. Medical History, 39, pp. 35–60

Quine, M.S. (2002) Italy's social revolution. Charity and welfare from liberalism to fascism. Basingstoke, Palgrave

Rapp, R.T. (1972) Review Alessandro Rossi e le origini dell'Italia industriale by L. Avagliano. Journal of Economic History, 32, pp. 945–946

Raseri, E. (1878) Il personale sanitario in Italia e all'estero. Studio comparativo. Annali di Statistica, 2, pp. 171–207

Raseri, E. (1883) Relazione sul movimento degli infermi degli ospedali civili. Atti del Consiglio Superiore di Statistica. Annali di Statistica, 7, pp. 123–130

Raseri, E. (1895) Notizie statistiche sull'epidemia di colera in Italia nell'anno 1893. Rivista di Igiene e Sanità pubblica, pp. 305–306

Riall, L. (1995) Legge marziale a Palermo: protesta popolare e rivolta nel 1886. Meridiana, 24, pp. 65–94

Robson N. (2008) Costing, funding and budgetary control in UK hospitals. A historical reflection. Journal of Accounting & Organizational Change, 4, pp. 343–362

Roe, D.A. (1973) A plague of corn. The social history of pellagra. Ithaca, Cornell University Press

Romano, R. (1982) Gli industriali e la prevenzione degli incidenti sul lavoro. In: Betri, M.L., Gigli Marchetti A., Eds. Salute e classi lavoratrici in Italia dall'Unità al Fascismo. Milano, F. Angeli, pp. 129–146

Rosenberg, C.E. (1966) Cholera in Nineteenth-Century Europe. A tool for social and economic analysis. Comparative Studies in Society and History, 8 (4), pp. 452–463

Rosenberg, C.E. (1979) Inward vision and outward glance. The shaping of the American hospital, 1880–1914. Bulletin of the History of Medicine, 53, pp. 346–391

Rosenberg, C.E. (1987) The care of strangers. The rise of America's hospital system. New York, Basic Books

Ross, R. (1923) Memoirs with a full account of the Great Malaria problem and its solution. London, John Murray
Rossi Doria, T. (1910) Il colera ai giorni nostri. Nuova Antologia, 234, pp. 151–157
Ruggi, G. (1924) Ricordi della mia vita. Bologna, Cappelli
Sallares, R. (2002) A history of malaria in ancient Italy. Oxford, Oxford University Press
Sambon, L.W. (1905) Remarks on the geographical distribution and etiology of pellagra. British Medical Journal, 2, pp. 1272–1275
Sanarelli, G. (1909) Dopo un trentennio di lotta contro la pellagra. IV Congresso Pellagrologico Italiano, Roma, Tip. Nazionale di Bertero & C.
Serao, M. (1884) Il ventre di Napoli. Milano, Treves
Silei, G. (2003) Lo Stato sociale in Italia. Storia e documenti. Vol. 1. Dall'Unità al Fascismo. Manduria, Lacaita
Sitta, P. (1899) Diffusione della pellagra in Italia. Giornale degli Economisti, 19, pp. 562–587
Sitta, P. (1900) La lotta contro la pellagra (i provvedimenti). Giornale deli Economisti, 20 (11), pp. 350–366
Snowden, F.M. (1991) Cholera in Barletta, 1910. Past & Present, 132, pp. 67–103
Snowden, F.M. (1995) Naples in the time of cholera, 1884–1911. Cambridge, Cambridge University Press
Snowden, F.M. (2003) Mosquitoes, Quinine and socialism of Italian women, 1900–1914. Past & Present, 178, pp. 176–209
Snowden (2006) The conquest of malaria. Italy, 1900–1962. New Haven, Yale University Press
Sormani, G. (1893) L'Igiene Pubblica ed il Socialismo. Rivista della Beneficenza Pubblica, 21, pp. 129–140
Stapleton, D.H. (2000) Internationalism and nationalism: the Rockefeller Foundation, public health and malaria in Italy, 1923–1951. Parassitologia, 42, pp. 127–134
Stevens, R. (1989) In sickness and in wealth. American hospitals in the Twentieth Century. New York, Basic Books
Sturdy, S., Cooter, R. (1998) Science, scientific management, and the transformation of Medicine in Britain, c. 1870–1950. History of Science, 36, pp. 421–466
Sydenstricker, V.F. (1958) The history of pellagra, its recognition as a disorder of nutrition and its conquest. American Journal of Clinical Nutrition, 6 (4), pp. 409–414
Tanzi, C. (1884) Notizie economico-statistiche sulla beneficenza e sull'ordinamento dell'Ospedale Maggiore di Milano e istituti annessi, 1874–1883. Milano, Tip. Cogliati
Thorne-Thorne, R. (1887) On sea-borne cholera. British measure of prevention v. European measures of restriction. BMJ, 2, pp. 339–340
Tognotti, E. (2000) Il mostro asiatico. Storia del colera in Italia. Bari, Laterza

Torelli, L. (1882) Carte della malaria dell'Italia illustrata. Firenze, Tip. Pellas
Tropeano, G. (1908) La clinica della malaria nel Mezzogiorno d'Italia. Napoli, Tip. Detken & Roll
Tropeano, G. (1912) La lotta contro il colera. Per l'educazione igienica popolare. Studi e Conferenze. Napoli, Tip. Detken & Roll
Verardini, F. (1873) Studi intorno al migliore ordinamento amministrativo e sanitario degli ospedali come richiesto dal progresso scientifico-pratico attuale. Annali Universali di Medicina, pp. 449–541
Villari, P. (1890) Nuovi tormenti e nuovi tormentati. Nuova Antologia, 30 (3), pp. 594–622
Vogel, M.J. (1980) The invention of the modern hospital. Boston, 1870–1930. Chicago, University of Chicago Press
Whitaker, E.D. (1992) Bread and work. Pellagra and economic transformation in turn-of-the-century Italy. Anthropological Quarterly, 65 (2), pp. 80–90
White Mario, J. (1895) The poor in great cities. Their problems and what is doing to solve them. New York, Scribner's Sons
Zocchi, P. Ed. (2008) La rete del perfezionamento medico. Milano scientifica, 1875–1924. Milano, Sironi
Zucchi, C. (1875) Sulle riforme desiderabili per il miglior ordinamento degli ospedali. In: Atti del VI Congresso dell'Associazione Medica Italiana tenuta in Bologna dal 22 al 28 Settembre 1874. Bologna
Zucchi, C. (1880) Sanità e beneficenza. Loro reciproci rapporti. Rivista della Beneficenza Pubblica e degli Istituti di Previdenza, 8, pp. 497–512
Zucchi, C. (1881) Del miglior governo degli spedali. Giornale della Società Italiana d'Igiene, 3, pp. 178–190
Zucchi, C. (1884) Il colera in Italia nel 1884. Giornale della Reale Società Italia d'Igiene, 9–10, pp. 507–523

CHAPTER 3

The Great War and the Spanish Flu

The period from 1915 to 1922, which saw the collapse of the liberal state and the rise of fascism, is crucial in Italy's social and political history. Central to this period is the Great War, which resulted in some 650,000 deaths and millions of disabled veterans, widows, and orphans. The disruption of the war reversed the progress against disease that had been achieved since the mid-1880s, exacerbated the impact of the influenza epidemic (the Spanish flu) which struck Italy at critical moments in the course of the war, between the spring of 1918 and the early months of 1919, and killed as many people as the war itself. A troubled postwar period of fierce social conflict, deep political polarization, and institutional paralysis hindered social and institutional reforms. This chapter examines the relationships among war, medicine, and the state, considering the impact of war on health in the Army and in the general population, the provision of health services with specific military import, focusing particularly on psychiatry and orthopaedics. It then considers the perplexing disappearance of sickness insurance from postwar plans. A special section is devoted to influenza, "the mother of all pandemics", whose management was limited by the state of medical knowledge and constrained by military logic and priorities, but which, in many ways, provides a preview of the mitigation strategies adopted with COVID-19.

Italy entered the First World War in May 1915, ten months after the beginning of hostilities, with the assumption of a short and victorious war that would provide the annexation of the *terre irredente* (un-redeemed

F. Taroni, *Health and Healthcare Policy in Italy since 1861*,
https://doi.org/10.1007/978-3-030-88731-5_3

territories) of Trento and Trieste, as negotiated with the new allies Britain and France in the secret London treaty. The Great War was a Total War, a laboratory (Gibelli 1982) and a workshop ("*officina*"—Gibelli 2014) which transformed the nation, accelerating processes already in operation and introducing new forces with lasting impact. Out of a population of about thirty-five million people, six million were called to arms. About 650,000 Italians died. Some 450,000 disabled veterans returned with serious injuries and over 200,000 with *malattie di guerra* (war diseases) such as tuberculosis and malaria. Millions of widows and orphans, demobilized soldiers, and women discharged from war-related industries along with refugees required support to readjust to civil and productive life. The tensions which were present at the beginning of the war intensified as the war dragged on. Instead of producing social cohesion and national solidarity, the Great War divided Italians in profound and long-lasting ways.

The belated decision to enter the war divided the nation in two politically heterogeneous camps, reflecting existing divisions manifested in the industrial and agricultural strikes and lockouts of 1913 and 1914, peaking with the *settimana rossa* (red week) of June 1914. The Interventionist front included the Nationalists pursuing the Great Patriotic War, which would build Italy's great power status, as well as Futurists, for whom the war was "the world's only hygiene", a *pharmakon* to get rid of both "the degenerate elements" of the Italian society and Giolitti's *Italietta* (small Italy) (Carter 2010, p. 69). The galaxy of "democratic interventionism" included Republicans, who saw the First World War as the fourth war of independence, and a host of prominent personalities such as Gaetano Salvemini, Giolitti's traditional archenemy, as well as former socialists such as Ivanoe Bonomi and Leonida Bissolati, a future Minister in Vittorio Emanuele Orlando's war cabinet. While Giolitti maintained a neutralist position, the anti-war stance was maintained principally by a composite bloc of the Left led by the Socialist Party, under the slogan "neither participation nor sabotage".

Throughout the war, the disastrous conduct of the conflict at the front and the harsh and repressive management of the home front provoked endemic social unrest. Popular protest against the spiralling cost of living and labour conflict in the militarized factories peaked with Turin's virtual insurrection in August 1917 (Procacci 1989), a few months before the military disaster of Caporetto. Commander in chief General Raffaele Cadorna took the defeat as a deliberate act of sabotage, "a military strike" organized by the "internal enemy" of "defeatists" and socialists (Procacci 1983, 2002). The collapse of the Italian front and the Austrian occupation of northeastern territories in October 1917 were a political and military watershed in the course of the war, which brought to a new government, a new commander in chief and a new strategy. However, the delusive victory of November 1918 further increased political polarization and social tensions. Although Italy was on the victorious side, it displayed social upheaval and political instability similar to the nations on the losing side. Gabriele D'Annunzio's myth of the *vittoria mutilata* (mutilated victory) and his occupation of Fiume mobilized Nationalists' sentiment against the Versailles' peace settlement, where Italy had obtained only part of the promised territories. Unmet expectations combined with a severe economic downturn of massive public debt at about thirteen million lire, hyperinflation, spiralling cost of living, and mass unemployment. A rising movement of peasants and industrial workers in what became to be known as *il biennio rosso* (the two red years) of 1919–1920 strengthened the Socialist Party, which won 120 parliamentary seats in the political election of 1919, 25 out of the 69 provinces and many of the major Italian cities, and culminated in September 1920 in the occupation of several northern factories (Lyttelton 1973). A militant political right also developed, which largely drew from veterans who glorified combat and scorned civilians, accused of dodging the military service (*imboscati*) and suspected of defeatism (Corner 2002).

Social radicalization combined with institutional instability. Seven different cabinets of short duration were formed between the end of the war and Mussolini's rise to power in 1922. The Parliament met rarely, and governments ruled by decree, provoking an institutional inertia that even Giovanni Giolitti, reinstated as prime minister in 1920, was not capable of overcoming. Many historians suggest that the origins of Fascism are to be found in the chaotic postwar situation that is usually defined as "the crisis of the Liberal State" (Corner 2002; see also a review in Carter 2011).

A Public Health Disaster

The war was a public health disaster that resulted in new needs and worsened existing health and social problems. The hardships of war and deteriorated hygienic conditions led to the resurgence of infectious diseases such as malaria and tuberculosis as well as cholera and smallpox in the Army and the civil population. The return of disabled soldiers imposed new medical demands far beyond the technical and financial capacity of beneficence and communal resources. In a reversal of the trend since the mid-1880s, total mortality rose in 1918 to 22.2 per 1000 population from 21.6 at the turn of the century, while life expectancy at birth fell to 30 years in males and 32 in females (Mortara 1925).

The resurgence of cholera in 1915–1916 spread the disease from troops occupying Austrian trenches of the Isonzo front among the Army, but was prevented from diffusing to the general population (Mortara 1925, pp. 272, 388). The rise of malaria instead affected both the Army and the civilian population. The state of war hindered the continuation of the anti-malarial campaign and the distribution of quinine, also used against the flu, and made drainage and sanitation impossible in the flood and marshes of the occupied territories of the northeast. In addition, about 80 per cent of the troops returning from Albania and Macedonia had contracted malaria. Malaria easily spread among the civilian population, where the increased number of women and children engaged in agricultural works exposed new subjects to infection. The large number of new malaria cases and the shortage of quinine reversed the steady decline in mortality observed since the turn of the century. Only by 1923 did malaria mortality return to its low levels of 1912.

Tuberculosis mortality also increased during the war, particularly in those regions, such as Lombardy and Tuscany, where it had been higher before. Compared to a death rate of 143 per 100,000 population in 1913, mortality from all forms of tuberculosis rose to 175 per 100,000 in 1917 and 209 per 100,000 in 1918 (Drolet 1945). Increased tuberculosis mortality was precipitated by the flu but was also associated with the rise of new cases in the civilian and the Army populations, where some 100,000 cases of "war tuberculosis" were estimated (Detti 1984). Stabilization of mortality occurred only in the 1920s. This however happened at higher rates than in the prewar period and with a large increase in pulmonary disease, which was covered in part by the continuing decline of other forms of tuberculosis (Mortara 1925, p. 376).

In addition to mortality from the flu, which will be discussed separately, the civil population experienced an increase in deaths from typhoid and diphtheria. The reversal of their prewar diminishing trends was associated with the general deterioration of hygienic conditions. Difficulty in access to medical care and serum therapy also contributed to rising diphtheria deaths. The Army was however relatively saved. Soldiers underwent several programmes of vaccination against smallpox, tetanus, cholera, and typhoid (De Filippi 1918). The dramatic fall of typhoid deaths in males from 17.9 per 100,000 in 1915 to 0.2 per 100,000 in 1919 and the increasing trend in the female to male ratio show the effect of inoculating large numbers of adult males in the Army. Its subsequent increase is associated with the gradual diminution of their relative immunity. This and the lower number of new cases of pellagra observed during the war are exemplars of what has been called "the goodness of war" to health and medicine (Cooter 1990).

The Medicalization of War

The essential function of medicine for the Army was to optimize the use of scarce human resources by reducing "waste" from wounds and diseases and promoting the prompt return of casualties to active service. The Great War was the first war in which more soldiers died of battle wounds than of disease. Eighty per cent of deaths in the Italian army resulted from enemy or friendly fire (Ferrajoli 1968). The fatality rate in severely wounded soldiers was 16.7 per cent, one-third lower than in the recent Libyan war (De Filippi 1918). The emerging life-saving powers of medicine combined progress in military surgery and standardization of the treatment of battle wounds (Selcer 2008), with prompt treatment close to the fighting line and rapid evacuation of wounded soldiers. The novel hierarchical organization of Army medical services differentiated the collecting zone from the battlefield, the evacuation and distributing zone, where the wounded were allocated to "territorial hospitals" to receive their final specialized treatment. Modern methods of rapid transport including mobile X-ray equipment, motorized ambulances, and trains (De Filippi 1918) were organized in collaboration with the Red Cross and with the support of American volunteers (Irwin 2009).

However, the urgent needs resulting from the conflict and the influenza epidemic resulted in a serious shortage of doctors. The Army enlisted a large proportion of civilian doctors and organized a *università castrense* (war university) delivering four-month courses to students in the last two years of their medical degree. By the end of the war there were only 874 regular *ufficiali medici* (medical officers) compared to the initial 770 and over 16,884 additional medical officers mobilized from civilian activities (Ferrajoli 1968). War and influenza strained the institutional fabric of traditional charity and put new burdens on *Opere pie* and communes. A 1917 hospital survey revealed a shortage of beds to meet local needs, let alone satisfy the demand for wounded and disabled soldiers (Pietravalle 1919). Hospitals subject to military requisition obtained military funding and the opportunity of upgrading their technologies, but this was a precursor of their postwar financial crisis. The transition to Army medical care with specialized technical skills and hospital resources drained civil institutions of professional manpower, left them in a state of distress and deterioration, and resulted in neglect of the competing needs of civilians, particularly in relation to the influenza epidemic. The shortage of doctors also hindered the development of some specialties that thrived in wartime in other countries, as predicted by the requirements for specialty development postulated in the classic work of George Rosen (1944). This was particularly the case for orthopaedics, "the paradigmatic specialty of the First World War" (Cooter 1990). In several countries, including Britain (Cooter 1993), the United States (Linker 2011), and Germany (Perry 2014), orthopaedics transformed, along with rehabilitation and vocational services, to help disabled soldiers' return to a full and productive life. In Italy, orthopaedics received large numbers of wounded and disabled soldiers to study and treat, and benefitted from significant technological innovations in the development of prostheses and artificial limbs, but suffered a dearth of specialized facilities and of human and financial resources and remained largely limited to a few centres in Milan and Bologna.

The Militarization of Psychiatry

Psychiatry was at the forefront of responding to military concerns over the optimal use of scarce human resources. It extensively supported the Army in determining the soldiers' capacity for physical and mental endurance, improving morale and manpower performance, and rationalizing military organization, including appropriate selection and optimal placement of conscripts. A relatively obscure medical subspecialty, psychiatry sought to improve its status while its practitioners took advantage of their commitment to the war effort to increase professional and social standing. A heterogeneous complex of psychiatrists, psychologists, anthropologists, and criminologists particularly of the Lombrosian tradition, best represented the concept of total war and the totalizing scope of the war culture applied to military and civilian activities. Placido Consiglio, a psychiatrist of the Lombroso school and captain of the Army, published a series of papers standardizing medical criteria to identify mental fitness at conscription, and the optimal placement of "the abnormal, degenerate and criminal" (Consiglio 1915–1916) to make "the best use of scarce human material". Agostino Gemelli, a friar, psychologist, medical captain, and consultant to the Supreme Command, analysed the qualities of the "good" soldier and studied "the causes predisposing to cowardice and heroism" (Gemelli 1917).

The militarization of medicine and the concomitant medicalization of war quickly expanded from managing the Army to governing the society at large. Placido Consiglio regarded the Army as a laboratory of applied eugenics where "an instructive form of social experimentalism" could be applied (Consiglio 1915). The principles of rational medical management initially developed "for the benefit of the army" were to be applied to the whole society to promote "the good of the social body" by antagonizing the reverse selection of "the worst" individuals and improve the nation's race (Consiglio 1916). Eugenics extrapolated from the military microsystem to the social macrocosm was understood as "a function of the state" administered in the first place by psychiatrists and physicians more generally. The First World War was a powerful catalyst for Italian eugenetics. The tools of "passive" eugenics of isolation and segregation to prevent the reproduction of the "undesirable" individuals and get rid of the "residuum" of society—the core of the Lombroso's theory of the criminal—implemented the Futurists' concept of the war as the hygiene of the world and imposed a regenerative turn on Eugenics. A heterogeneous ideological melange of a wide spectrum of social, anthropological, and eugenics doctrines combining progressivism, Darwinism, and even socialism voiced a discourse of the dysgenic or "degenerogenous" consequences of the war and the pressing needs of the biologic regeneration of the nation in the quality and quantity of its population (Cassata 2006, p. 52; Mantovani 2003, 2004, p. 201). The rhetoric of efficiency and the power of the nation were a precursor to Fascist theories of the New Men (Cassata 2011) at both the individual and population levels (Gini 1912).

The assumed modernizing role of medicine, and of psychiatry in particular, in war and society put medical doctors at the forefront of the national elite and made "the ruling of the Nation… the duty of medical doctors in present times" according to Enrico Morselli, one of the most important leaders of the psychiatric movement (Scartabellati 2008). The self-attributed leading role of medical doctors based on the medicalization of war was mirrored by the militarization of medicine. In the aftermath of the "Manifesto of the Ninety-Three" German Intellectuals supporting the German Army military actions and of the ensuing breakdown of the concept of universality and communality of science, Italian psychiatry overturned its traditional epistemological and clinical relationship with the German school (Babini 2009, p. 49). Morselli's journal *Quaderni di*

Psichiatria published a series of inflammatory editorials and papers explaining with anthropological and biological arguments the differences between the Latin genius and the German inferior civilization, and even reproaching alleged Germans' psychiatric disturbances and the purported degeneration of Kaiser Wilhelm himself (see, e.g., Anonymous 1917).

Disabled Soldiers and Social Reform

Wars and their aftermath have been traditionally associated with social and institutional reforms responding to the social needs created or made manifest by the conflict. The warfare-to-welfare nexus, that is, the view that national solidarity created by the common experience of war is a catalyst of social reform and the development of the welfare state, is now a disputed concept (Obinger and Petersen 2015. For an extensive review of the warfare-welfare nexus in Italy, see also Ferrera 2018). The debate particularly applies to postwar Italy, where a number of social programmes to deal with the legacies of war were put forward but quickly floundered.

Legislation for disabled veterans and support for widows and orphans were considered early in the course of war, as soon as the Chamber reconvened in December 1916 after a six-month recess. Aiming at a veterans' swift return to productive civil life, the bill envisaged a plan where disabled soldiers were required to stay for no less than three months (which were part of military service) in specialized hospitals providing medical, physical, and vocational services as well as prostheses and artificial limbs. Rehabilitation programmes were to be managed by the military, traditional charities and purposeful beneficent organizations, and financed by the communes. Specialized hospitals required a proper hospital design and equipment as well as a number of medical and professional specialists. However, parliamentary debates made clear that the number of beds, widely planned between 2000 and 4000, were well below the need of the number of disabled soldiers, running from about 10,000 to 15,000 but possibly more than twice those numbers. Moreover, orthopaedic services had been a small and relatively obscure subspecialty in the prewar years and specialists were scarce, with only four University chairs in orthopaedics. The government opposed a National Institute of Rehabilitation and repeatedly rejected proposals to expand the number of chairs of orthopaedics and even to make the teaching of orthopaedics obligatory for obtaining the medical degree.

The state escaped direct intervention in a typical Giolittian mode. To remedy the shortage of specialized beds the state offered long-term loans to communes, which they frequently declined to avoid future expenses on day charges. Moreover, governance was devolved to *Opera Nazionale Invalidi di Guerra* (ONIG-National Organization of War Disabled), a national confederation of a loose network of public, private organizations and local spontaneous Committees (Detti 1984). ONIG was autonomous from the state and was a precursor to the practice of delegating decision-making from ministerial to technical, "parallel bureaucracies" (Melis 1988) characterizing the *Enti* (Agencies) of the so-called *para-stato*, which mushroomed during the Fascist regime (Salvati 2006).

A veterans' rehabilitation law was not passed by the Senate until mid-1917 and did not achieve much in terms of returning disabled soldiers to autonomous and active life. The law however was acclaimed by the socialist deputy Angiolo Casalini as "the first example of the new social legislation born out of the war". The socialist plan was to force further social reforms along the traditional path of the *assicurazioni operaie* as a solution to *questione sociale*. Socialist deputies first tried to supplement rehabilitation services to productive work with veterans' pensions and then to exploit financial provisions for disabled veterans to bridge the transition towards a comprehensive social insurance package for the whole population. The bridging legislation focused in the main on soldiers' pensions for *tubercolosi di guerra* (war tuberculosis) (Detti 1984). The assumption was that since the experience of war had led to the wide acceptance of the idea that "combating tuberculosis is a legitimate function of the state" (Maffi cited in Detti 1984, p. 887), then state assistance would easily be expanded further to cover sickness along with all "other normal ailments of life". Leonida Bissolati, a former socialist, and after Caporetto Minister of *Assistenza Militare e Pensioni di Guerra* planned to transform it into a *Ministero della Assistenza e della Previdenza*, responsible for a comprehensive scheme of social insurance over old age and disability, work accidents, unemployment, and sickness. This would oversee "**all** forms of assistance benefitting **all** those affected by the war" (emphasis in the original) transforming the state into the so-called *Stato Giuridico-Sociale* (Ministero per L'Assistenza Militare e le Pensioni di guerra. L'Assistenza di Guerra in Italia 1919, p. 31). A less radical proposal was for a Ministry of Hygiene and Public Health to bring public health and assistance to the poor under the same roof, unifying the various agencies spread across several Ministries, which had caused so many tensions during the war (Pietravalle 1919). While Britain instituted a Ministry of Health in 1919 and France a Ministry of Hygiene, Aid and Social Pensions in 1923, in Italy both proposals failed miserably. Francesco S. Nitti first downgraded the Ministry of Military Assistance and War Pensions into a new directorate of the Treasury, and then instituted a Ministry of Labor and Social Providence, from which sickness insurance was excluded. This conformed Nitti's productivist orientation and reasserted the lavorist orientation of welfare which characterized prewar Giolitti's policies.

While legislation extending workmen compensation to agricultural workers was passed in 1917 and the laws for compulsory social insurance against old age and unemployment were approved in 1919 (Pavan 2019), sickness insurance never materialized. An extensive proposal for a quasi-universal system of sickness insurance was developed by *Commissione Abbiate*, which was established in August 1917 and reported rather quickly in 1919, pragmatically envisioning a "minimalist" and a "maximalist" version to account for financial constraints (Giorgi and Pavan 2021, pp. 79–85). The final report of the *Commissione Abbiate* was given the blessing of the *Commissione Reale per il Dopoguerra*, particularly in its Section on *Legislazione sociale e previdenza*, which was responsible for planning the transition to peace time. Its design summarized much of the prewar debate on the compulsory nature and tripartite funding of *assicurazioni operaie*, was accepted by organized doctors, including *medici condotti* and hospital practitioners, and by political parties of liberal, catholic, and socialist orientation. It also promised to bring relief to hospital budgets providing them the enduring source of finance that hospital doctors had dreamed of in 1911. However, the project was never discussed in Parliament and just evaporated from the political agenda.

The goal of using veterans' pensions as a bridge to the social insurance of tuberculosis and then to sickness insurance was overly ambitious both practically and politically. It underestimated the technical and social difficulties of transforming a paternalist programme focused on a narrowly defined group of worthy veterans into a broad-based social insurance scheme oriented to a much wider group of citizens, based on need rather than worthiness and merit. As opposed to the warfare-to-welfare nexus, social cohesion and the sense of national solidarity were scarce resources in postwar Italy riven by political polarization and social conflicts. Moreover, instead of a situation of "war socialism" in terms of increased fiscal and administrative capacity, Italy experienced a postwar period of institutional paralysis with a series of extremely short and weak governments, while hyperinflation and a massive public debt resulting from the cost of waging the war left little room for increasing public expenditure and taxation. More generally, detailed plans for reform developed by a host of technical commissions clashed with what was widely perceived as a pre-revolutionary situation of the society, exhibiting a deep divide between the "real" and the "legal" country. Some have argued that the Great War took Italy "from a rear guard position to one at the forefront" in terms of welfare legislation (Ferrera et al. 2012, p. 45). In matters of health and sickness insurance, however, the Italian system remained trapped in the no-man's land between the traditional liberal schemes of poor relief of Crispian and Giolittian ancestry and the new forms of liberal-democratic provision which took off in various countries in the aftermath of the Great War. The Fascist regime brought to completion and implemented most of the missed innovations debated in the chaotic years of the crisis of the Liberal State.

Influenza Epidemic and the War

The influenza epidemic of 1918–1919 was "the mother of all pandemics" (Taubenberger and Morens 2006) causing from 30 to 100 million deaths worldwide according to the various estimates (Johnson and Mueller 2002), an impressive number even at the time of the COVID-19 pandemic. Italy was the European country hardest hit by the epidemic, with five to six million cases, about 10 per cent of the population, and 274,041 officially reported deaths (Istat 1958, p. 229) compared to about 600,000 excess deaths between August 1918 and March 1919 estimated by Mortara (1925, p. 380). The oft-repeated statement that influenza led to about the same number of deaths as the First World War seems therefore to apply to Italy.

In hindsight, the course of the epidemic is traditionally described as a compact period of crisis, subdivided into three distinct waves. The relatively benign first wave in the spring and summer of 1918 was followed by a more virulent strike in autumn, peaking at the end of October, after which a milder third wave occurred in the first months of 1919. During the springs of 1920, 1922, and 1923, the Army registered an upsurge of cases, which raised debates about whether these were new waves or a resurgence of the epidemic (Mortara 1925). At the time however experts were baffled by the striking difference in severity between the deadly autumnal strike and the early cases (defined *mal gentile*—gentle disease—and "the three day fever"), as much as their relationships with the very mild flu epidemic of 1889–1890, with only 12,292 official deaths (Istat 1958). The new events were generally taken as an entirely new disease, often of exotic origin, and it was only later that the different waves came to be seen as a single condition with a common aetiology (Bresalier 2011).

The epidemic hit the southern regions first and showed wide regional variability in mortality, with the highest levels observed in Lazio and the lowest in Veneto, at about 12 and 5 deaths per 1000 population respectively (Mortara 1925, p. 381). Two-thirds of the victims were young adults aged twenty to forty years, the opposite of traditional influenza epidemics, and predominantly female, possibly because they were more exposed because of participating the war mobilization efforts in the home front.

Italy shared worldwide medical uncertainties over the origins, the aetiology, and the treatment of the disease, which were the subject of fierce debates between clinicians and bacteriologists of different schools (Tognotti 2002, 2003). The controversy had symbolic, clinical, and epistemological aspects and consequences. To characterize the disease as the well-known influenza or "grippe" was thought to diminish the fearsome nature of the epidemic and keep the population calm in wartime. An urgent *circolare* (executive order) transmitted to the Prefects by Prime Minister V.E. Orlando stressed that the *Consiglio Superiore di Sanità* had certified that since the disease was "only" influenza, and not "an exotic and deadly disease", it was a much less serious threat than perceived (cited in Tognotti 2002). Uncertainty over the aetiology of the disease led to confused clinical recommendations for the treatment of the disease and the public health management of the epidemics. Doctors experienced "a sense of bewilderment and dismay" (Dragotti 1918) and recommended either strictly conservative ("absolute rest and as few drugs as possible") or aggressive treatments with quinine, camphor, and other cardiotonics, aspirin, opium and antitussives, colloidal gold, and so on. A host of vaccines and sera against various pathogens were also used, including "auto-vaccines" developed from the same patient. "Polytherapies" produced a general shortage of drugs, particularly quinine, which was distributed to the troops against malaria, and up to a sixfold increase in their prices, which were the focus of popular anger among doctors and pharmacists (Tognotti 2002, p. 101). In Italy, the use of vaccines and sera from Istituto Sieroterapico Milanese followed the general frenzy strongly criticized by a Lancet editorial (Lancet 1918) and the American Public Health Association, which developed a special working programme against its excesses (APH 1919a). Both lamented that research was "hurried and incomplete" and resulted in unjustified claims of efficacy, which even leading journals accepted too easily. To dispel "the fog of research" (Tomkins 1992) the American Public Health Association responded setting explicit standards for research on vaccines and sera and evaluating their effectiveness in its "Working program against influenza" (APH 1919b).

Public health authorities relied on a broad array of interventions, which combined specific measures against the air-born infection with generic public health interventions (Giovannini 1987). These included anti-crowding measures; the closure of public gathering places such as churches, cinemas, and theatres; and the ban on funerals and weddings; and disinfection with menthol and phenol of public places such as railway stations. Individual protection focused on personal hygiene and disinfection of laundry and beddings, as well as the mouth, nose, and throat. Wearing gauze masks was also recommended but rarely applied, even among health professionals. National provisions issued by the *Direzione Generale di sanità* and the War Ministry were variously integrated by local municipal departments, particularly from the major cities such as Milan (see, e.g., Corriere della Sera September 28). The measures to mitigate the spread of influenza were very similar to those adopted in France (Hildreth 1991; Rasmussen 2010) and by several American cities (Markel et al. 2007; Tomes 2010). An exception was "the stoic attitude" officially adopted by the British government at the national level (Honingsbaum 2013). Arthur Newsholme, Chief Medical Officer of the Local Government Board exposed the British strategy to "just carry on" at the emergency meeting of medical authorities held at the Royal Society of Medicine on November 13, 1918, a few days after the armistice and at the peak of the epidemic. In his famous "Discussion on influenza", Newsholme explained the reasons of the decision to shelve the mitigation programme he had prepared and exhorted to "just carry on". His main motivations were the dearth of knowledge ("I know of no public health measures which can resist the progress of pandemic influenza", p. 3), and a sense of patriotic duty in times of war ("the relentless needs of warfare justified incurring this risk of spreading infection") that defined "national circumstances in which the major duty is to carry on" (Newsholme 1918).

Newsholme's motivations raise the issue of how waging war and influenza interacted in Italy. Influenza struck when the population was weakened by three years of war and at two critical moments in its course, the River Piave resistance to the Austrian offensive in May 1918 and the Allied counter-offensive with the battle of Vittorio Veneto in October. War-related strategic aims, such as the movement of troops and medical efforts, took precedence over civilian concerns, generating tensions between military and civil administrations. The competition for hospital beds and doctors was intense, as about 80 per cent of doctors and pharmacists had been mobilized and the dearth of *medici condotti* was frequently lamented by the population (Tognotti 2002, p. 144). Moreover, tensions arose between the goal of convincing the population to follow the government's *Istruzioni popolari* by taking the dangers of influenza seriously and adjusting their behaviour accordingly and the logic of war propaganda to minimize risk and avoid panicking the population. Moreover, the influenza epidemics and its uncertainties had shaken the faith in bacteriology and the progress of scientific medicine. An editorial of Nuova Antologia exposed how "medicine is not making a good show" in dealing with "a disease so silly and yet so terrible" (Anonymous 1919, p. 234), and Enrico Bertarelli welcomed "a banal and quasi-comic disease to punish our pride" (Bertarelli 1918). Newspapers stigmatized conflicting medical practices as "doctors' *confusionismo*" (Pacchioni 1919). After years of dazzling successes, bacteriologists had to confront a bitter failure, as they were seen "always doing research but never reach a consensus" (Pontano 1918). At stake was the Pasteurian paradigm that the mastery of the disease required isolating the germ and then developing a serum or a vaccine (Eyler 2010). To the *Riunione Italiana per lo studio della pandemia* held in Milan to take stock of the epidemic participated the great clinicians of the time, including Luigi Devoto and Luigi Mangiagalli. All raised merciless criticisms against bacteriology and praised the virtue of differential diagnosis and the clinical approach (Patriarca and Clerici 2018). The voice of Bartolomeo Gosio, head of the Laboratory of Bacteriology of the Direzione generale di sanità and a strenuous defender of the Pfeiffer's bacterium *Haemophilus influenzae* as the cause of influenza, remained isolated. He wished that the final proof of the aetiology of influenza could be found "in the peace of the laboratory during a pause from the frenzy times, which often require hurried announcements" (Gosio 1922, p. 7). This however was simply beyond the scope of current medical knowledge, as it was not until 1933 that a British research team eventually identified the virus of influenza (Bresalier 2011).

References

Anonymous (1917) Sullo stato mentale dei popoli belligeranti. Quaderni di Psichiatria,4, pp. 224–225

Anonymous (1919) Contro l'influenza. Nuova Antologia, 284, p. 234

APH—American Public Health Association (1919a) A working program against influenza. American Journal of Public Health, 9 (1), pp. 1–13

APH—American Public Health Association (1919b) Weapons against Influenza. American Journal of Public Health, 8, pp. 787–788

Babini, V.P. (2009) Liberi tutti. Manicomi e psichiatri in Italia: una storia del Novecento. Bologna, Il Mulino

Bertarelli, E. (1918) Il male del giorno. Rivista d'Italia, 3 (2), p. 228.

Bresalier, M. (2011) Uses of a pandemic. Forging the identities of influenza and virus research in interwar Britain. Social History of Medicine, 25, pp. 400–424

Carter, N. (2010) Modern Italy in historical perspective. London, Bloomsbury Academic

Carter, N. (2011) Rethinking the Italian Liberal State. Bulletin of Italian Politics, 3, pp. 225–245

Cassata, F. (2006) Molti, sani e forti. L'eugenetica in Italia. Torino, Bollati Boringhieri

Cassata, F. (2011) Building the New Man. Eugenies, racial science and genetics in twentieth-century Italy. CEUP Press

Consiglio, P. (1915) Studi di psichiatria militare. Parte Quarta. Proposte e rimedi. Rivista Sperimentale di Freniatria, 41, pp. 35–80

Consiglio, P. (1915–1916) I militari anormali in guerra. Rivista di Antropologia, 20, pp. 3–15

Consiglio, P. (1916) La rigenerazione fisica e morale della razza mediante l'esercito. Rivista Militare, 61, pp. 23–51

Cooter, R. (1990) Medicine and the goodness of war. Canadian Bulletin History of Medicine, 12, pp. 147–159

Cooter, R (1993) Orthopedics and the organization of modern medicine, 1880–1948. London, Palgrave Macmillan

Corner, P. (2002) The road to Fascism: An Italian Sonderweg? Contemporary European History, 11 (2), pp. 273–295

De Filippi, F. (1918) On some special problems of the Italian medical war services. Journal of the Royal Society of Medicine, 11, pp. 40–60

Detti, T. (1984) Stato, Guerra e tubercolosi, 1915–1922. In. Storia d'Italia. Annali, Vol. VII, Malattie e Medicina. Della Peruta F. Ed. Torino, Einaudi, pp. 877–951

Dragotti, G. (1918) Polmonite influenzale e polmonite pestosa. Il Policlinico, 25 (51), pp. 65–69

Drolet, G.J. (1945) World War I and tuberculosis. American Journal of Public Health, 35, pp. 689–697

Editorial (1918) The utilization of vaccine for the prevention and treatment of influenza. Lancet, ii, p. 565

Eyler (2010) The state of science, microbiology and vaccine circa 1918. Public Health Reports, 125, pp. 27–36

Ferrajoli, F. (1968) Il Servizio Sanitario Militare nella Guerra 1915–1918. Giornale di Medicina Militare, 118 (6), pp. 501–516

Ferrera, M., Fargion, V., Jessoula, M. (2012) Alle radici del welfare all'italiana. Origini e sviluppo futuro di un modello sociale squilibrato. Venezia, Marsilio
Ferrera, M. (2018) Italy: Wars, political extremism and the constraints to welfare reform. In: Obinger, H., Petersen, K., Starke, P., Eds. Warfare and welfare. Military conflict and welfare state development in Western countries. Oxford, Oxford University Press, pp. 99–126
Gemelli, A. (1917) Il nostro soldato. Saggi di psicologia militare. Milano, Treves
Gibelli, A. (1982) La guerra laboratorio. Eserciti e igiene sociale verso la guerra totale. Movimento Operaio e Socialista, 3, pp. 335–349
Gibelli, A. (2014) L'officina della Guerra. La Grande Guerra e le trasformazioni del mondo mentale. Torino, Bollati Boringhieri
Gini C. (1912) I fattori demografici nella evoluzione delle Nazioni. Torino, F.lli Bocca
Giorgi, C., Pavan, I. (2021) Storia dello Stato Sociale in Italia. Bologna, Il Mulino
Giovannini, P. (1987) L'influenza "spagnola". Controllo istituzionale e reazioni popolari, 1918–1919. In: Pastore, A., Sorcinelli, P. Eds. Sanità e Società, Udine, Casamassima, pp. 373–398
Gosio, B. (1922) Ricerche batteriologiche sull'influenza. Introduzione. Annali di Igiene, pp. 2–17
Hildreth, M.L. (1991) The influenza epidemic of 1918–1919 in France. Contemporary concepts of aetiology, therapy and prevention. Social History of Medicine, 4, pp. 277–294
Honingsbaum, M. (2013) Regulating the 1918–1919 pandemic. Flu, stoicism and the Northcliffe press. Medical History, 57 (2), pp. 165–185
Irwin, J.F. (2009) Nation building and rebuilding. The American Red Cross in Italy during the Great War. The Journal of the Gilded Age and Progressive Era, 3, pp. 407–439
ISTAT—Istituto Centrale di Statistica. (1958) Cause di Morte, 1887–1955. Roma
Johnson, N., Mueller, M. (2002) Updating the accounts. Global mortality of the 1918–1920 "spanish" influenza pandemic. Bulletin of the History of Medicine, 76 (1), pp. 105–115
Linker, B. (2011) War's waste. Rehabilitation in World War I America. Chicago, Chicago University Press
Lyttelton, A. (1973) The seizure of power. Fascism in Italy, 1919–1929. London, Routledge
Mantovani, C. (2003) Rigenerare la stirpe. Il movimento eugenetico italiano e la Grande Guerra, 1915–1924. Ricerche di Storia Politica, 2, pp. 203–224
Mantovani, C. (2004) Rigenerare la società. L'eugenetica in Italia dalle origini ottocentesche agli anni trenta. Soveria Mannelli, Rubbettino
Markel, H., Lipman, H.B., Navarro, J.A., Sloan, A., Michalsen, J.R., Stern, A.M., Cetron, M.S. (2007) Nonpharmaceutical interventions implemented by US cities during the 1918–1919 influenza pandemic. Jama, 298 (6), pp. 644–654

Melis, G. (1988) Due modelli di amministrazione fra liberalismo e fascismo. Burocrazie tradizionali e nuovi apparati. Roma, Ufficio centrale per i beni archivistici

Ministero per l'Assistenza Militare e le Pensioni di Guerra (1919). L'assistenza di guerra in Italia. Roma, Poligrafica Italiana

Mortara, G. (1925) La salute pubblica in Italia durante e dopo la guerra. Bari, Laterza

Newsholme, A. (1918) Discussion on Influenza. Proceedings of the Royal Society of Medicine, 12, pp. 1–102

Obinger, H., Petersen, K. (2015) Mass warfare and the welfare state, causal mechanisms and effects. British Journal of Political Science, 47, pp. 203–227

Pacchioni, D. (1919) È o non è l'influenza? Pathologica, 245, pp. 60–64

Patriarca, C., Clerici, C.A. (2018) Pathologica ai tempi della Spagnola. Pathologica, 110 (4), pp. 316–320

Pavan, I. (2019) War and the Welfare State. The case of Italy, from WWI to Fascism. Historia Contemporanea, 61 (3), pp. 835–872

Perry, H. (2014) Recycling the disabled. Army, medicine and modernity in WWI Germany. Manchester, Manchester University Press

Pietravalle, M. (1919) Per un Ministero della Sanità ed Assistenza pubblica in Italia. Nuova Antologia, 284, pp. 103–117

Procacci, G. (1983) Stato e Classe operaia in Italia durante la Prima Guerra Mondiale. Milano, F. Angeli

Procacci, G. (1989) Popular protest and labour conflict in Italy, 1915–1918. Social History, 14 (1), pp. 31–58

Procacci, G. (2002) The disaster of Caporetto. In: Disastro! Disasters in Italy since 1860. Culture, politics and society. Dickie, J., Foot, J., Snowden, F.M., Eds. London, Palgrave, pp. 141–161

Pontano, T. (1918) Note cliniche, epidemiologiche ed eziologiche sull'attuale epidemia di influenza. Il Policlinico, 40, pp. 10–18

Rasmussen, A. (2010) Prevent or heal, laissez-faire or coerce? The public health politics of influenza in France, 1918–1919. In: Giles-Vernick, T., Craddock, S. Eds, Influenza and Public Health. Learning from Past Pandemics, Londra, Earthscan, pp. 69–83.

Rosen, G. (1944) The specialization of medicine with particular reference to ophthalmology. Springfield, Thomas

Salvati, M. (2006) The long history of corporatism in Italy. A question of culture or economics? Contemporary European History, 15 (2), pp. 223–244

Scartabellati, A. (2008) Il dovere dei medici italiani nell'ora presente. Biopolitica, seduzione bellica e battaglie culturali nelle scienze umane durante il primo conflitto mondiale. Medicina & Storia, 14, pp. 65–94

Selcer, P. (2008) Standardizing wounds. Alexis Carrel and the scientific management of life in the First World War. British Journal of the History of Science, 41 (1), pp. 73–107

Taubenberger, J.K., Morens, D.M. (2006) 1918 influenza. The mother of all pandemics. Emerging Infectious Diseases, 12 (1), pp. 15–22

Tognotti, E. (2002) La “Spagnola” in Italia. Storia dell’influenza che fece temere la fine del mondo (1918–1919). Milano, F. Angeli

Tognotti, E. (2003) Scientific triumphalism and learning from facts. Bacteriology and the “Spanish flu”, challenge of 1918. Social History of Medicine, 16, pp. 97–110

Tomes, N. (2010) Destroyer and teacher. Managing the masses during the 1918–1919 influenza pandemic. Public Health Reports, 125, pp. 48–62

Tomkins, S.M. (1992) The failure of expertise. Public health policy in Britain during the 1918–1919 influenza epidemic. Social History of Medicine, 5 (3), pp. 435–454

CHAPTER 4

Health Under the Fascist State

The chapter provides a profile of the complex set of policies and institutions of the fascist regime, which ruled Italy from 1922 to 1943. The classic periodization of the regime envisages its seizure of power, consolidation, and expansion (Paxton 1998, 2004). This provides the context for the emergence of health policies with distinct fascist characteristics in the late 1920s, when power was centrally concentrated in the state (statization) and was seized by the *Partito Nazionale Fascista* (PNF—National Fascist Party) (fascistization), and their full implementation throughout the 1930s. The early 1930s saw the consolidation of the fascist social state (Quine 2002) and the building of its main institutions in the midst of the Great Depression, which in Italy peaked in 1932. Expansion occurred during the economic uptake of the second half of the 1930s, which is associated with the darkest period of the regime. This is characterized by the war in Ethiopia and the ensuing politics of self-sufficiency (autarchy) and the closer alliance with Nazi Germany, starting with participation in the Spanish Civil War (1936) and the Rome-Berlin Axis, formalized by the Pact of Steel in 1939. This led to Italy's entry in the Second World War in June 1940 and precipitated the fall of the dictatorship in July 1943.

The fascist regime prided itself as the most original and inclusive system of social services worldwide. Benito Mussolini's bold statement that "fascist social legislation is the most advanced in the world" embellished the masthead of *Politica Sociale*, the leading political and ideological review in social policy. At the core of fascist social policy were two documents, the

F. Taroni, *Health and Healthcare Policy in Italy since 1861*,
https://doi.org/10.1007/978-3-030-88731-5_4

Carta del Lavoro (Labor Charter) and the *Discorso dell'Ascensione* (Ascension Day address) issued in April and May 1927 respectively. Neither of them was a legislative act but both exerted a significant and long-lasting ideological influence as well as a strong practical impact.

The Labor Charter was a collection of thirty short *dichiarazioni* (statements) of the *Gran Consiglio del Fascismo*, the regime's supreme political body, defining the principles and aims of Corporatism, the institutional order, and political economy of the fascist state that embraced labour and capital under the direction of the state. Three statements envisaged various social insurance schemes including general health insurance, with insurance against tuberculosis and for some occupational diseases as the first steps.

Benito Mussolini's Ascension Day address to the chamber of the deputies was a call for political mobilization, which laid out a far-reaching programme for the regeneration of the nation's social body with the goal of increasing the number and strength of the population. To achieve "maximum natality, minimum mortality" and ensure the nation's self-reliance, the Duce called for "a battle for Births" and "a battle for Wheat", and identified himself as "the doctor of the Nation" who took care to cure its scourges. The different logics of the two pillars of fascist health policy is an example of the syncretism of fascist policies, which some scholars characterize as "a ragbag" of ambiguous and contradictory ideas (Luzzatto 1999). The regime addressed many of the social issues traditionally neglected or only belatedly addressed and incompletely dealt with by the liberal state, frequently adopting and adapting institutions first invented in the liberal era. Also, several health and social institutions developed under fascist rule remained as structural features of the Republican welfare State. These backward and forward continuities explain why even in the health and social field fascism cannot be isolated as a "parenthesis" in the political and institutional life of the country and how the historical analysis of its performance and alleged "achievements" is still a sensitive topic in Italy.

The Ascension Day Address: A Political and Ideological Agenda

Benito Mussolini's address to the Chamber of Deputies of May 26, 1927 (the day of Jesus Christ's ascension to heaven in the Catholic liturgy), set down the ideological principles and laid out a policy agenda, which consolidated activities under way and future initiatives into a long-term programme. The overarching goal was to attain a numerous and healthy population, which would accomplish the quest for economic self-reliance and satisfy the nation's imperial aspirations. As Mussolini argued in his preface to the Italian translation of Richard Korherr's book, it was not the quality but the quantity of the population which mattered most as "the strength of the Nation is in its number" (*il numero fa la forza*) (Mussolini 1928).

Mussolini pointed to the traditional ills of malaria, tuberculosis, and syphilis, as well as the new problems of cancer, alcoholism, and suicides as warning signals of degeneration and national decline, of which however the diminishing birth rate was the surest sign and the biggest menace. In Mussolini's view, degeneration was the effect of a "destructive urbanization" that "sterilized the people". Citing the big cities of Turin and Milan where the number of deaths nearly surpassed births, city life was described as bringing moral decay, neurosis, and a consumer culture attracting foreign goods, which run counter to the austere, prolific, and self-sufficient model of life of the rural population. Italians had to be made "*molti, sani e forti*" (many, healthy, and strong) by transforming their public and private lives and mobilizing popular sentiments as well as medical science (Cassata 2006). The goal was not just to "make Italians" as the common trope of Risorgimento suggested, but to re-make them and create the "New Man" (Bernhard and Klinkhammer 2017), and the new woman as well. Fascist propaganda represented women as the agents and the victims of the urban degeneration. The urban *donna-crisi* (crisis woman) was cosmopolitan, fashionable, skinny, hysterical, and sterile, compared to the rural *donna-madre* (woman mother), who was instead nationalistic, florid, calm, and prolific (de Grazia 1992, p. 6; Chang 2015).

The strong pro-natalist policy of the Ascension Day address had a long-standing ideological and political impact upon the regime's social and health policy. Its concepts dovetailed with widespread concerns and remedies, which had emerged in various countries before and after the First World War. Fears of degeneration in decadent industrializing and modernizing societies had been a recurrent theme in *fien-de-siecle* European circles (Pick 1989). Eugenics had progressively emerged as the technocratic instrument to manage the population and secure national efficiency. In France, the vision of a healthy nation had particularly stressed the protection of mothers and children, making eugenics part of social medicine and public health (Schneider 1991). In Italy, eugenics had fascinated professional and political elites, from nationalists to socialists (Mantovani 2004, p. 201), and "the development of a strong population" dominated the vision of the *Commissionissima's* Report on the priorities of postwar Italy.

The real watershed with previous liberal policy was the claim for the direct and comprehensive ("totalitarian" in common fascist parlance) intervention of the state, overturning the Liberal "suicidal theory" that "the State should not be concerned with the health of its population" predicated on the theory of laissez-faire. Turning from his own short "Manchesterian" phase, Mussolini now predicated "in a well-ordered State, population health is the utmost priority" and therefore medicine had a prominent position as "a function of the State". Fascist pro-natalist policy was largely based on two different scientific, quantitative eugenic paradigms: the "integral demography" of the demographer and statistician Corrado Gini based on the theory of the cyclical rise and fall of populations (1912) and the endocrinologist Nicola Pende's "orthogenesis" (Cassata 2006, p. 135). Pende's constitutionalist biotypology aimed at constructing the New Man through his science of "political biology" or orthogenesis, that is, by correcting the habits and neuro-endocrinologic milieu of individuals who deviated from their optimal biotype (Beccalossi 2020).

On the practical side, *Opera Nazionale Maternità e Infanzia* (ONMI—National Organization for Mothers and Children), in Mussolini's words *"un'organizzazione fascistissima"* (the most fascist of organizations), aimed at instilling women with a sense of motherhood as a patriotic duty to the nation and providing pre- and perinatal care to mothers and children (see also p. 108). The fascist policy of "totalitarian demography" (Ipsen 1996) went well beyond medical science and practice and progressively reached all spheres of collective and private life. Restrictions were imposed upon migrations both to the cities, to slow down the degenerating process of urbanization, and overseas, now seen more as a loss of valuable national resources than a gain of precious remittances. The new penal code of 1931 classified abortion as "a crime against the stock" (*stirpe*, a term initially preferred to *razza*, or race) and prosecuted contraceptive practices. Bachelors and childless couples were taxed, and the revenues allegedly accrued to ONMI. Positive measures focused on mobilizing rural women through the party mass organization of *Massaie Rurali* (Rural Housewives-Wilson 2002) and promoting large families with a wide array of tax breaks, bonuses, loans and prizes on marriage, births and large families delivered in local and national propagandistic rituals (de Grazia 1992, p. 70).

Furthermore, rejecting Malthusian theories, Mussolini's parliamentary speech called to sustain a larger population by increasing productivity of available land and "redeeming" the swamplands infested with malaria to additional cultivable areas. This energized the *battaglia del grano* officially launched in 1925 with the economic goal of attaining self-sufficiency in wheat production in order to limit dependency on imported grain and help the balance of payment and expanded it into the ambitious project of *bonifica integrale* (integral land reclamation), which was closely tied to the battle for birth and the ruralization campaign (De Felice 1974, p. 142).

The Ascension Day address provided an ideological and rational imprint to the most important "battles" to which the regime called the Italian population. The battles' rhetoric of "perpetual struggle" instilled a sense of urgency, which demanded total commitment and continuous devotion as a patriotic duty and allowed inconsistencies, contradictions, and frequent reformulations of the processes and goals of the various programmes (Nelson 1991). The initiation of the programme of *bonifica integrale* in the early 1930s was a showcase of the achievements of fascism, the demonstration of its entrepreneurship and modernizing drive. Mussolini's image driving tractors, ploughing soil, harvesting, and threshing wheat bare-breasted as a sign of virility and power was ubiquitous in newspapers, documentaries, and films, and greatly contributed to his personal cult (Falasca-Zamponi 1997).

Bonifica Integrale: Land Reclamation and Anti-malarial Policy

Land reclamation schemes were not new to Italy as hydraulic and drainage works went back to Roman times, and an important programme of land reclamation was launched with Baccarini's law of 1882 (Bevilacqua and Rossi Doria 1984). The idea that land reclamation should also include comprehensive agrarian transformation emerged in the aftermath of the First World War from a community of agricultural economists, engineers, agronomists, and jurists of which Arrigo Serpieri was the prominent figure (D'Antone 1979). Serpieri designed a complex strategy of hydraulic drainage combined with infrastructural works (building canals, ditches, and roads) and intensive agricultural transformation (mechanization, chemical fertilization, and high-yield seeds). Moreover, new juridical tools such as *consorzi di bonifica* (land reclamation consortia) would ensure bridging hydraulic reclamation with agricultural transformation and assuring the financial co-participation of landowners and local public administrations

(Fumian 1983). As Undersecretary for Land Reclamation up to the mid-1930s, Arrigo Serpieri passed the first legislation for *Grande Bonifica* (Great Reclamation) in 1923 and 1925. His pioneering laws were overshadowed by Mussolini's law of December 1928, the only legislation that ever assumed his denomination, which institutionalized the concept of *Bonifica Integrale* (integral land reclamation) as "the effective coordination of three forms of *bonifica*—hydraulic, agricultural and hygienic—through the cooperation of local officers and private bodies under the direction of the State" (Buccella 1929). *Bonifica igienica* was only a small, but essential, part of the much broader project of social engineering, and medicine did not play the leading role it had in the early 1900s' anti-malarial vertical programmes. Nevertheless, lasting progress required that the breeding grounds of malaria were cleared, and the permanent settlers of the reclaimed land were protected.

Out of the seven billion liras over a fourteen-year period committed to the programme, most went to the advanced sectors of the north, but attention focused on the Pontine Marshes, a vast and depopulated marshland in the southern Latium, not far from Rome and the focus of several unsuccessful projects beginning in Roman times. The internal colonization of the Pontine Marshes involved about 60,000 peasants and their families from Veneto, Friuli, and Romagna (Caprotti 2008), and was organized by *Opera Nazionale Combattenti* (ONC-National Veterans' Organization) founded in 1917 and closely linked to the Fascist Party (Barone 1984). ONC was entrusted with expropriating the reclaimed land and allocating it to the peasant families. Settlers were share-tenants of the allotted farms (*poderi*), which they could purchase from ONC with 15-year instalments. Farms were organized into 18 rural centres (*borghi*) bearing the names of battle-sites of the First World War. Five modernistic New Towns (*Città Nuove*) were built in record time, from 1932 to 1939 embodying the symbol of a modernizing plan for the entire Nation (Caprotti 2007), the first step in "the reclamation of the Nation's stock" through internal migration (Pende 1933, p. 241).

Land sanitation and health protection of the settled population were organized on the strategies and tools devised by the prewar anti-malarial campaigns of the Giolittian era. Quininization (now using the new synthetic drugs) of sick people adopted Grassi's recommended approach and was based on the dispensaries and rural health stations first experimented with by Angelo Celli (see p. 46). The network of rural health centres, dispensaries, and children's sanatoria maintained relations with the two traditional Roman "fever hospitals", Santo Spirito and San Giovanni. Coordination was assigned to a *Comitato Provinciale Antimalarico* (Provincial Antimalarial Committee) introduced in 1934 (Snowden 2006, p. 142). The Rockefeller Foundation, which was active in Italy beginning in 1923 with Lewis Hackett, supported programmes against mosquitoes and their larvae and treated pools and stagnant waters with Paris green and larvivorous fishes (Stapleton 2000).

As a social engineering project transforming a vast and depopulated marshland into a dynamic and productive community, the *bonifica integrale* of the Pontine Marshes was widely publicized in Italy and attracted significant attention abroad. It was "the best known, most publicized, and most discussed of the works carried out by Fascism" (Dogliani 1999, p. 202; see also the extensive discussion in De Felice's analysis of "the years of consensus"—De Felice 1974) both in Italy and abroad, and particularly in the United States (see, e.g., Schmidt, 1937), where parallels were drawn with the Tennessee Valley project of the New Deal America (Ghirardo 1989; Schivelbush 2006). However, as a general strategy against malaria, the scientific community received *bonifica integrale* with scepticism. In the late 1920s malariologists were divided between those favouring anti-mosquitos measures (a strategy defined species-sanitation or *piccola bonifica*—small reclamation—in Italy) and advocates of general social improvement, while both camps marginalized the palliative role of quinine for sick people (Evans 1989).

The 1925 Report of the survey of the League of Nations' Malaria Commission concerning the efficacy of various methods of malaria control recommended adopting "various, multiple, local and simple measures" with the important provision that existing tools were not "yet capable of eradicating the disease" (League of Nations. Malaria Commission 1925, see also 1939; Gachelin 2013). Nicholas Swellengrebel, an outstanding Dutch entomologist, argued that there was no convincing evidence on the effectiveness of *bonifica integrale* as an anti-malarial measure: except hydraulic works, not a single element of the complex Italian programme had a direct effect on malaria endemicity. Distinguishing "the hygienic value" of *bonifica integrale* from its "economic merits" in promoting growth and improving sanitary conditions, the Malaria Commission concluded that the Italian project could not be recommended as a model for malaria control (League of Nations. Malaria Commission 1925, p. 70). Scepticism was also expressed by the larger scientific community convened in Rome by the League of Nations for the First International Congress on Malariology of October 1925. Mussolini's inaugural address pointed to the great scientific and social merits of the Italian policy of *Grande Bonifica* (Mussolini 1925), and participants visited several sites of *bonifica*. However, the delegates dodged the resolution of the Congress submitted by Gaspare Gosio, director of the microbiologic laboratory of the *Direzione generale di sanità*, that called *Grande Bonifica* "a real and significant step to combat malaria" (cited in Gachelin et al. 2018). To the Congress, Lewis Hackett reported instead on his experiments in Sardinia with larvicides, which he considered the only effective tool to control malaria (Hackett 1925). He later found the adoption of "general measures of social uplift and hygiene" in the fight against malaria "a confession of ignorance or defeat" and dismissed quinine of "little use as a public health measure" (Hackett 1937, p. 268).

Hackett and the Rockefeller Foundation maintained the use of larvicides for mosquito control as "an article of faith" of anti-malarial policies up to the introduction of DDT (Stapleton 2000). The sceptical attitude of the League of Nations towards *bonifica integrale* instead changed in a few years (Gachelin et al. 2018). The 1927 Report of the Malaria Commission recommended the control of parasites in men by treatment with quinine and the destruction of mosquitos in houses, and recognized that a general improvement in sanitation was more effective than mosquito control alone in lowering the incidence of malaria. The 1939 European Conference on Malaria's Report explicitly acknowledged *bonifica integrale* as the most effective long-term strategy for malaria control (League of Nations 1939). In Italy, victory in the Pontine Marshes was officially claimed in December 1939 at the *Mostra Nazionale delle Bonifiche* (National Exhibition of Land Reclamation). However, attempts to transplant the model to Apulia and Campania were interrupted by the war, and these programmes were never brought to completion (Snowden 2006, p. 165).

To assess the achievements of the battle for wheat and its essential tool *bonifica integrale* is not an easy task considering its complexity, the regime's hyperbolic rhetoric, and the dubious reliability of contemporary statistical sources. Scholars agree that the battle for wheat was successful in terms of increasing production and reducing wheat imports (Cohen 1979). Moreover, it helped in spreading mechanization and the adoption of more rational and intensive methods of cultivation such as deep ploughing, seed selection, and use of chemical fertilizers, and intensified research particularly in plant genetics (Saraiva 2011). This, however, was achieved at the cost of serious imbalances in other sectors of the agricultural economy such as wine, oil, and animal husbandry, and did not contribute to improving conditions for peasants as protectionist tariffs increased food prices. Statistics from the new Provincia of Littoria (now Latina) showed a clear success in malaria control, with a striking decline in the frequency of cases

(Snowden 2006, p. 172). These statistics of remarkable health achievements did not include cases in the occasional workers of the early *bonifica idraulica* and should be put in the context of the declining trend of malaria deaths observed in Italy since the turn of the century.

The Institutional Pillar: Carta del Lavoro and Casse Mutue

Carta del Lavoro (Labor Charter) was the other pillar of fascist health policy, the basis of its approach to health insurance against tuberculosis, occupational diseases and other so-called *malattie comuni* (common disease conditions). The Charter's thirty brief statements laid down the principles of fascist corporatism, the new political order alternative to both capitalism and socialism based on the integration of the competing interests of labour and capital under state guidance for the betterment of the nation (Bottai and Turati 1929). The Charter introduced the concept of "solidarity between productive factors… in the higher interest of productivity" (Statement IV) as a form of cross-class collaboration, which stemmed from the duty to participate in the national effort to increase productivity. Tracing the path to "national solidarity" Mussolini explained how fascist social insurance differed from nineteenth-century mutual aid societies, which were "instruments of class struggle" based on "social hate", as well as from charity and philanthropy, manifestations of piety, and compassion of the riches towards the poor (Mussolini 1931). "*Mutualità sindacale*" (syndicalist mutuality) through *casse mutue di malattia* (mutual sickness funds) were the tools of the new solidarity and the embodiment of the corporatist doctrine in health insurance.

Three statements defined *casse mutue* as the "unshakable monuments to corporatism" (Pisenti 1930). Labour contracts underwritten by "recognized syndicates" would set up *casse mutue* "when they were technically feasible" (Statement XXVII), and employers and employees would bear "a proportionate contribution" in their funding (Statement XXVI) and participate in their management "with the supervision of professional organization and corporate bodies" (Statement XXVIII). Statement XXVII laid down an agenda envisaging, among several other forms of social insurance, "insurance against tuberculosis and occupational diseases as a first step towards a general insurance against all diseases".

Funding of social and health insurance was left to employers' and employees' "proportionate contributions" with a view towards minimizing the state's financial stake, in continuity with liberal financial policies. The parenthetical but fundamental clause that labour contracts would establish *casse mutue* only "when technically feasible" left their institution to the occasional convergence of the goodwill of fascist syndicates (the only form of unionism that the state accepted and "recognized") and of the industrialists' organization of the relevant corporate economic sector. While insurance for tuberculosis and six occupational diseases was passed in a few months (although the latter went into operation only five years later), the Industrialists' Confederation (*Confindustria*) resisted introducing binding social insurance in labour contracts. In 1929, the Supreme Council for the National Economy deferred compulsory general health insurance and encouraged instead the development of separate funds for specific occupational categories (Giorgi and Pavan 2021, p. 153). This explains the chaotic and uneven diffusion of *casse mutue* as well as their wide variation in terms of benefits and contributions.

The dearth of data makes a detailed profile of the diffusion of *casse mutue* in the four sectors of the corporate economy difficult to describe except in general terms. The industrial sector came first and spread quickly, particularly in the north. The 1107 *mutue* which were active in 1929 almost doubled to 1978 in 1938. The relatively affluent sector of commerce and credit followed, while *casse mutue agricole* for the traditionally neglected class of peasants appeared only later. Official sources show that in 1934, the 1978 *casse mutue dell'Industria* (industry sickness funds) had about 1.4 million insured compared to the just seven *casse agricole* with about 120,000 insured (PNF 1936, p. 60). The bulk of agricultural workers was covered only in 1937, when sickness insurance extended to farm tenants and sharecroppers, their most affluent categories. Giorgi and Pavan report that in 1939–1940, 6.4 million persons were covered in agriculture, 7.4 million in industry, 382,000 in commerce, and 242,000 in credit. Moreover, 18.5 million were insured against tuberculosis (Giorgi and Pavan 2021, pp. 158–159).

The insured population included *semi-abbienti*, wage earners of moderate means and from the lower middle class. As some of them were also included in the municipal list of the poor entitled to free medical care, the poor list became the uncertain boundary between municipal assistance and health insurance. All *mutue* covered medical services for domiciliary visits and hospitalization, but some did not include drugs for the policyholder and/or for his relatives; most provided sick payments of about half or two-thirds of current wages for 90 or 180 days per year. Except *casse commercianti* (shopkeepers), which refunded medical expenditure incurred by their insured, *mutue* relied on closed lists of contracted doctors for service provision. Contract practices were not exclusive, and most doctors contracted with several *casse* and many practised also as *medici condotti*. Form of payment differed by occupational sector: sickness funds of the industry paid their doctors fee-for-service, agricultural funds a lump sum per capita, and commerce refunded patients of their expenses, the method of payment doctor preferred. Form of payment and fee schedules were negotiated between the fascist medical syndicate and *casse mutue*, which were controlled by the syndicate of the relevant corporative sector. Workers' contributions had a ceiling of about 2 per cent of their wages, widely assumed much lower than the insurance's financial break-even point. This contributed to the financial weakness of *casse mutue* along with their fragmentation into thousands of small, local agencies with wide variation in benefits and contributions.

The chaotic organizational and financial structure of health insurance explains the unrelenting and unsuccessful attempts at integrating and even merging *casse mutue*. For example, when the four National Confederations of *casse mutue* for the corporatist sectors of industry, trade, agriculture, and credit and banking were created, their financial management was kept separate, and this frustrated the goals of financial stability and efficiency (Angelini 1937). Equally unsuccessful was the attempt in 1939 to curb the number of *casse mutue* by raising their minimum threshold from 100 to 700 insured. Eventually the January 11, 1943, law merged *casse mutue* into a single *Ente Mutualità Fascista. Istituto per l'Assistenza Malattia ai Lavoratori* (Mutual Fascist Organization. Institute for workers' disease assistance). However, this was only a few months before the dissolution of the regime on July 25, too late to patch up one of the purported achievements of the regime, but timely enough to institutionalize a path conditioning the health system of the Republic for decades. Entering the medical marketplace, *casse mutue* challenged the autonomy of the medical profession as well as its earnings, and this radically changed both community and hospital medical practice (see below).

A World Apart: The Battle Against Tuberculosis

The passing of the 1927 compulsory insurance legislation against tuberculosis was a watershed, the transition from voluntary and charitable initiatives towards a state-directed, centrally funded policy, including a massive programme of sanatoria building. Here again, the Regime took over an issue which had been neglected, or unfinished, by the Liberal state. Although tuberculosis was the principal cause of death resulting in about 60,000 deaths per year, Giolittian policy had maintained that the proper function of the state was only in addressing its causes, while patient assistance had to be left to charities and municipalities (Giolitti Chamber March 17, 1904). Occasional initiatives flourished, when and where local finances permitted. The philanthropic initiative of the city of Milan opened the first "popular sanatoria" for middle-class patients in 1903 in Valtellina, in the Alps. A survey by the Italian Federation of local voluntary anti-tuberculosis organizations in 1914 found only twenty-three dispensaries and six sanatoria, of which five were "popolari" with 484 beds in total (Cherubini 1974). In the aftermath of the First World War, which saw a sustained surge of tuberculosis, and after the failed attempt to introduce specific insurance against "war tuberculosis", state policy adopted the traditional scheme of low-cost, long-term loans guaranteed by the state, which municipalities traditionally refused to accept, afraid of the burden on future expenditures. Moreover, the Commission for Tuberculosis, instituted with the prompting of the Rockefeller Foundation and the American Red Cross, in Italy failed to participate in building sanatoria and dispensaries as it did in France (Murard and Zylberman 1987). It was only influential in opening schools of nursing in various cities, including visiting nurses (*assistenti sanitarie visitatrici*) for home visiting in dispensaries and maternal and child services (Fiorilli 2016).

The law for compulsory insurance against tuberculosis was passed in 1927 and entered into force in January 1929 with a target of about twenty million insured out of a total population of about forty million. Coverage included dependent workers between fifteen and sixty-five years of age and clerks earning less than 800 lire per month (the same population covered by old age and disability insurance) and was extended to spouses and children of fifteen years or less. By 1929, 8.5 million workers were insured, and total coverage was about 18 million people. In 1936 insurance was extended to about three million land tenants and sharecroppers, and in 1938 to elementary schoolteachers raising the insured population to 18.5 million.

Funding was provided by *Cassa nazionale delle Assicurazioni Sociali* (CNAS), which in 1933 was transformed into the *Istituto Nazionale Fascista della Previdenza Sociale* (INFPS—Fascist National Institute of Social Providence), which also managed and funded social insurances for disability and old age, maternity, and unemployment. The insured risk did not cover the entire clinical course of the disease but was limited to tuberculosis *in atto* (under way) later defined *attiva* (active). Obvious clinical uncertainties in determining the insured risk led to continuous friction between INFPS, *casse mutue*, and their doctors over the attribution of clinical and financial responsibilities. Benefits included residential accommodation in sanatoria or in specialized wards of general hospitals, and admission refusal or voluntary discharge implied the loss of all benefits. Children and spouses were paid a small grant during sanatorium stay of the insured; they expired in six months but were renewable in case of readmission or longer stays. Home and ambulatory care was specifically excluded from coverage, except "as an exceptional and temporary accommodation" in the early years when beds in sanatoria or specialized hospitals were not yet available (Francioni 1930a).

In the interwar years, tuberculosis could be easily diagnosed through X-rays and sputum samples, and its community spread was contained to some extent through sanitary measures. In terms of cure, instead, medicine could only try to increase patient resistance to the progression of the disease through fresh air, bed rest, and diet regimens. However, obtaining a permanent halt of the disease progression was difficult and long, and required the full collaboration of patients who frequently resisted admission, the long stay and the unpleasant conditions and strict regulations of sanatoria. The claim for improving housing conditions and particularly overcrowding as a means of tuberculosis control was a common view among the professional community, frequently voiced in professional journals, such as *Lotta contro la Tubercolosi* (Fight Against Tuberculosis), official organ of the *Lega Italiana contro la Tubercolosi* (Italian League Against Tuberculosis) and its meetings' proceeding (see a review in Preti 1987). However, as Eugenio Morelli made clear "the solution of the housing problem is a matter of billions of liras, while building sanatoria is a matter of million liras" (Morelli 1943).

The regime decided to invest part of the financial reserves derived from workers' contributions to tuberculosis insurance in the construction and operation of sanatoria. As estimates were that of the total 60,000 beds needed (current assumption was that one bed was needed for each tuberculosis death) only about one-fourth were available in sanatoria or, more frequently, in special hospitals, or dedicated wards of general hospitals, the tuberculosis insurance law provided for a ten-year plan for 20,000 more beds. In charge of the plan was *Ufficio Costruzioni Sanatoriali* (Sanatorial Building Committee) of INFPS, dominated by Eugenio Morelli, a specialist in pneumonology and tisiology and an influent party-loyalist named *il Duce dei sanatori*, member of the Chamber of the Deputies and then appointed to the Senate, for a long-time secretary of the doctors' syndicate. The office quickly devised detailed architectural plans and operational standards for sanatoria in different locations and exposed to various climates (the plain, the seaside, the hills, the high mountains) and, in the words of Eugenio Morelli, transformed Italy into "a gigantic construction site" (Del Curto 2010). Figure 4.1 shows a general profile of the progress in building sanatoria beds across the years 1929–1940.

By the end of 1939, INFPS owned and managed 49 sanatori with a capacity of over 16,000 beds (Francioni 1957). The type and size of structures dedicated to the assistance of patients with tuberculosis at different stages varied enormously. The most renowned hospital in Rome was initially entitled to Benito Mussolini, who dedicated it instead to Carlo Forlanini, who pioneered the use of pneumothorax, hosted over 2000 beds, with a vast array of specialized wards for all types of tuberculosis and a variety of surgical services. The Villaggio sanatoriale di Sondalo dedicated to Eugenio Morelli was a high-altitude sanatorium in the Alps, the largest sanatorium in Europe with over 3000 beds. More of a real village than a mere hospital, its construction, which included scraping the mountain to create space for the buildings, started in 1932 and was concluded only in the 1950s, at an astounding cost of 88 million liras (Del Curto 2010). A dearth of data allows only very rough estimates of the cost of the sanatoria project. Projecting over the ten-year duration of the programme the detailed information available for the year 1937, Preti estimates total cost at about five billion lire, of which two-thirds was from the INFPS budget (Preti 1987, p. 189).

In contrast with the luxurious programme of sanatoria building, the institution of dispensaries followed a troubled path. *Dispensari antitubercolari* were the operating arms of *Consorzi Provinciali Antitubercolari* (CPA—Antitubercular Provincial Consortia) originally created in 1917, but legislated again in 1927 when they were administered and funded by provincie and their communes on a per-capita basis. Their statutory functions were limited to defining diagnosis, case finding and contact tracing in the family and the community (Francioni 1930b). However, as "an exceptional and temporary case", until specialized beds were available,

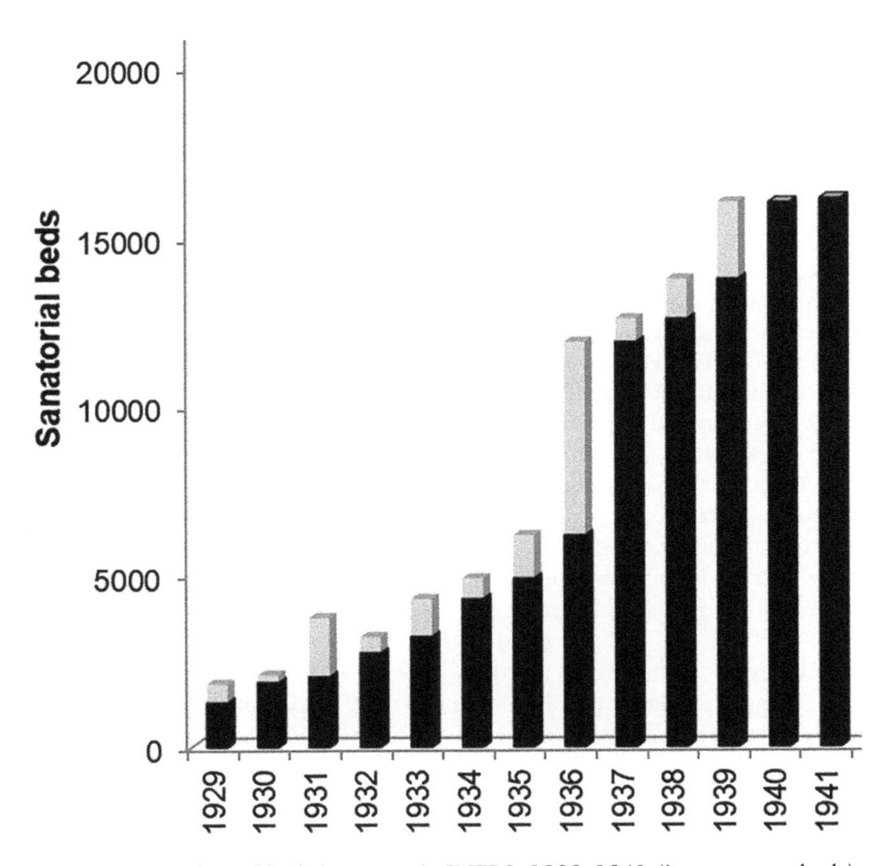

Fig. 4.1 Number of beds in sanatoria INFPS, 1929–1941 (in grey: new beds). (*Source*: Francioni, 1957)

they also had to provide ambulatory and home care for patients and their families, in addition to organize home visits and after-care through *assistenti sanitarie visitatrici* (visiting nurses), as well as provide educational activities and health promotion and information (Roatta 1938). In practice, dispensaries usually run very tight budgets and were poorly equipped in terms of X-rays and staff, which enjoyed little professional status. Their most compelling activity was to allocate uninsured patients to sanatoria, for which they received a meagre additional grant from the state.

Against traditional "totalitarian" claims, the most striking characteristic of fascist policy against tuberculosis was its separation from general sickness insurance and its fragmentation into a dual system, which was split in terms of insurance coverage, funding, and management, which severed central funding for residential accommodation in sanatoria from local contributions for dispensaries' community orientation. In Britain, the Sanatorium Benefit of the National Insurance Act (NIA) also provided for the construction of sanatoria, but general health insurance covered both tuberculosis and other "common" diseases, which in Italy were instead covered by *casse mutue*. Moreover, the "British scheme" included both sanatorium and dispensary care for all members of the community under the responsibility of local County and County Borough Councils (Bryder 1988). In Italy, the lack of integration between the various agencies involved in the care of tuberculosis in its different stages compromised early diagnosis and the prompt admission of patients, further reducing the little good that sanatorium treatment could do. Uncertainties over the benefit of sanatoria in "curing" tuberculosis or halting the progress of the disease had started very early, when Robert Koch's Nobel lecture in 1905 mentioned the "relatively small number of real cures obtained in the sanatoria" (Koch 1905). Sanatoria isolated infected persons, limiting the community spread of the infection, but their only tools in the interwar years were fresh air and bed rest on sleeping porches and verandas, a rich diet, and hygienic instructions. Artificial pneumothorax, introduced by the Italian pneumologist Carlo Forlanini in 1888, as with other more complex and riskier surgical "collapse therapies" of the lung did not clearly improve the therapeutic impact of sanatoria. Flurin Condrau has shown that from one-third to one-half of patients died within five years after discharge from sanatoria, most relapsed and were readmitted several times, and only one-third returned to productive work (Condrau 2010). Reports from the dispensaries show similar data in Italy. Patients were often admitted to sanatoria at advanced stages of the disease, which implied long lengths of stay and increased mortality, both in hospital and after discharge. Estimates from CPAs were that most patients relapsed within one year, and up to about two-thirds died within five years from discharge (see, e.g., Gualdi and D'Argenio 1940; see also the review in Preti 1987, pp. 182–185). Moreover, admission to sanatoria was not easy, particularly for patients who were not insured against tuberculosis. The 1942 annual statistical yearbook of the Federazione Nazionale Fascista per la Lotta alla Tubercolosi shows that out of 70,029 cases with tuberculosis variously traced in 1938, only 42,926, or about 60 per cent, were admitted to a sanatorium or some specialized facility (Giorgi and Pavan 2021, p. 172).

Based on admittedly very incomplete data, one could conclude that the sanatoria programme absorbed an estimated five billion liras to accommodate about two-thirds of detected cases and lose up to two-thirds of their allegedly "restored" patients within five years of their discharge. The apparent debacle did not raise a debate over the benefits of relying on sanatoria, as the Ministry of Health did in Britain in the 1930s (see, e.g., Bryden 1988, p. 200) or about an alternative, "home treatment" of tuberculosis, which emerged in the United States in the postwar years and intensified with the economic constraints of the Great Depression. Pointing to the controversies about the clinical effectiveness of sanatoria, Lynda Bryder in Britain (1988) and Eva Eylers (2014) in Germany have focused on their additional functions as instruments of political legitimation and social control. In Italy, the sanatorium crusade can also be interpreted as propaganda in the international political arena and a mechanism of national politics and control. The centralization of the battle against tuberculosis and the establishment by INFPS of a dense network of local offices under strict party control consolidated the influence of the centre over the periphery and was an important opportunity for political patronage at the local level (Giorgi 2004).

The Fragmented Fascist Healthcare State and Its Doctors

The regime intensified the consolidation of its expanding welfare state in the early 1930s while moving towards a new "totalitarian" phase of "organic co-ordination" with distinct characteristics of "staticization" and "fascistization" (Quine 2002, p. 96; Giorgi 2012). The consolidated structure of the fascist healthcare state had three main constituent parts: the new welfare-based *Enti pubblici*, traditional ministerial bureaucracies, and local services and administrations, albeit in a much-diminished role. The new structure, which was legislated into *Testo Unico delle Leggi Sanitarie* (Integrated code of sanitary laws) in 1934, did not include *casse mutue*, which notionally pertained to the Ministry of Corporations. They however radically transformed the healthcare market, changed doctors' practice, and hospital activity.

Enti pubblici were specialized public-sector agencies enjoying vast administrative and financial independence under state supervision. Mariuccia Salvati distinguishes *Enti Previdenziali*, such as INFPS and

INFAIL, which managed and funded social insurance from those concerned with finance (IMI—*Istituto Mobiliare Italiano*) and industry (IRI-*Istituto per la Ricostruzione Italiana*) created in 1932 and 1933 to facilitate money supply and accelerate industrial recovery from the Great Depression (Salvati 2006). Welfare-based *Enti* were more numerous and varied in size, financial power, and political influence. The two biggest were INFPS and INFAIL, both answering to the Ministry of Corporations and funding and managing respectively the main programmes of social insurance and the insurance against injuries and occupational diseases. The third of the biggest *Enti, Ente Mutualità Fascista* was instituted in the twilight of the Regime and did not contribute to the functioning of the Fascist welfare system.

Out of the several smaller and often slightly legally different organizations, ONMI was the most relevant for the health field, the primary agency for the implementation of the populationist health policy with the goal of coordinating the provision of services for mothers and children. ONMI was a national confederation of over 5000 public and private local organizations, run and staffed by lay and medical volunteers and heavily dependent on private and municipal funding (Quine 2002, p. 129). ONMI combined the political mission of promoting motherhood as a patriotic duty and the modernizing project of standardizing and promoting modern techniques of "rational motherhood" and childrearing (Garvin 2015). Compared to finance-based *Enti*, which maintained a tradition of technical expertise frequently based on personal relations with Mussolini, the activity of *Enti* in the field of health and social care was intensely politicized. Chiara Giorgi has clearly documented how INFPS was led centrally by party-loyalists and developed a vast network of local branches deeply ingrained in local affairs, acting as political conduits between centre and periphery and instrumental in orchestrating consensus and organizing control (Giorgi 2004). Service expansion was pursued through a dense stratification of the entitlements of various categories of beneficiaries, who were subject to cumbersome procedures of access, which amplified administrative discretion to discriminate against "unworthy" applicants. The diffusion of *Enti*'s "parallel bureaucracies" to those of the Ministries therefore brought modernization in the management and provision of services but also greatly helped the centralization and "fascistization" of health and social organizations.

Despite the image of decisiveness, efficiency, and bureaucratic rationality that the regime tried to convey, the expansion in scope and range and the centralization of "totalitarian" programmes entailed fragmentation and competition both between the two parallel bureaucracies of *Enti* and the Ministries and within the Ministries themselves. INFPS administered and funded tuberculosis insurance in competition with municipal services and sickness insurance administered by *casse mutue*, although it was eventually answerable to the Ministry of the Economy, and then of the Corporations, as *casse mutue* also were. The Ministries of the National Economy and later of Corporations, of Agriculture, of Public Works and the Ministry of the Interior through its *Direzione di Sanità Pubblica* had a stake in land reclamation programmes, whose management in the Pontine Marshes was entrusted to *Organizzazione Nazionale Combattenti*, itself a special *Ente pubblico*. Reporting on the Italian malarial campaign in 1932, Hackett described the effects of this competing bureaucracies in terms of "paralyzing turf wars" between the Ministry of Public Works and the Ministry of the Interior's Department of Public Health. The latter allegedly lack of credit in political circles and of technical leadership with public health services resulted in a maze of "mutually independent, inconsistent and often ineffective malarial activities scattered all over the country" (Hackett cited in Snowden 2006, p. 179).

The centralized nature of fascist health policy eroded the multiple health powers accrued by municipal governments during the Liberal era. The political eclipse of municipalities during fascism has several and complex causes (Melis 1996, p. 345). In the health field, municipalities were simply shut out of the significant innovations of the interwar years. This is particularly apparent when compared with the golden age of Britain's municipal services (Gorsky 2011). Maternal and child services, an element of modernization of community services, were the responsibility of the local branches of ONMI; sanatoria were owned by INFPS and managed by its local branches under strict central supervision, while dispensaries, their community counterpart, were governed by provincial CPA and developed apart from municipal offices of hygiene. The 1934 *Testo Unico* of health laws transferred the responsibility of microbiological and chemical laboratories from *comuni* to *provincie*, while *medico provinciale*, appointed by the Prefect, stood above *ufficiale sanitario*, contracted by *comune*. Furthermore, municipal administration of domiciliary and hospital services for the poor competed with *casse mutue*, which provided the same services through often the same doctors.

Casse mutue were the most significant long-term innovation introduced by the fascist regime. They were a fragmented, unequal, and inefficient system of health insurance with a fragile financial base. However, their establishment was a watershed in the structure and functioning of the healthcare marketplace which upset the practice of private practitioners, *medici condotti*, and hospital doctors. Moreover, *casse mutue* by opening the hospitals to non-poor, insured patients further contributed to "hospitalize medicine" and to its specialization. The only hospital census performed in the *ventennio* (twenty years) of fascist rule shows that the increased number of hospitals and their expanded social function did not result in significant changes in their geographical distribution (Istituto Centrale di Statistica 1934). In 1932 there were 2090 public and private hospitals, twice as many those existing in 1907, with over 200,000 beds, or 5.7 per 1000 population (Table 4.1). Hospitals continued to be concentrated in the big cities of the northern part of the country, with a range between 8.6 beds per 1000 population in Lombardy and Tuscany and 0.5 and 1.2 beds per 1000 in Lucania and Calabria respectively. Average length of hospital stay was 59 days, ranging from 17 in surgical (mostly private) hospitals and 30 days in general hospitals to 259 days in *manicomi* (mental hospitals). Most private hospitals (*Case di Cura* or *Case della Salute*) were small facilities specialized in general surgery, gynaecology (excluding maternity), and eye diseases. General hospitals were mostly public and accounted for three-quarters of beds and admissions, but the hospital census also shows a heavy specialist bias in the distribution of

Table 4.1 Public and private hospitals by type, 1932

Hospitals	*Nr.*	*%*	*% private*	*Beds %*	*ALOS*[a]
General	1319	63.1	20.8	44.8	30
Surgical	215	10.3	39.2	2.0	17
Mental	154	7.4	10.1	32.7	259
Tuberculosis	154	7.4	12.4	11.0	121
Chronic	83	4.0	0.8	4.6	222
Gynaecological	36	1.7	6.3	0.3	12
Paediatric	30	1.4	0.8	1.4	37
Eye	23	1.1	2.9	0.4	24
Orthopaedics	15	0.7	0.8	0.6	36
Isolation	13	0.6	0.0	1.0	28
Other specialized	48	2.3	5.9	1.2	n.a.
Total hospitals	2090	100.0	100.0	239,009	59

Sources: Istituto Centrale di Statistica, Statistica degli ospedali, 1934

[a]Average Length of stay, days

hospitals and their beds. The high number of tuberculosis beds, both in sanatoria and in dedicated beds of general hospitals, reflects the priorities of the time. The high proportion of tuberculosis beds and admissions reflects both the sanatoria programme and the transformation of general hospital beds incentivized by the reliable and rich source of funding from INFPS for patients insured against tuberculosis. Also of note is the number of surgical hospitals, mostly privately owned, which took advantage of the new lines of business opened by self-paying patients and doctors' interest in private practice. The high proportion of beds in mental hospitals was the effect of the massive programme of new *manicomi*, usually very large and concentrated particularly in the south, to accommodate the increased number of admissions and, particularly, their long stay, as discharge had to be authorized by the judiciary. This imbalance created absurd situations, which at times reached the floor of the Parliament, such as the case of Palermo, which had a general hospital capacity of 300 beds and was furnished with a luxurious new *manicomio* of about 3000 beds (Chamber 9 April 1932).

Hospital doctors were twice harmed by the increased number of admissions of insured patients, who required their more continuous presence in hospital wards and deprived them of the income from self-paying clients. This was also important for private practitioners and *medici condotti* providing home care. Patients now insured with *casse mutue* were workers of moderate means usually not included in the municipal list of the poor entitled to free medical care, who traditionally provided the main source of private, self-paying patients for hospital doctors as well as private practitioners and *medici condotti*. The coming of *casse mutue* therefore deprived private practitioners of their living, and *medici condotti* of additional income, unless doctors accepted mutual funds contracts, which they perceived as cheap and threatening professional autonomy and self-governance.

Maria Malatesta explains how the new situation found Italian doctors and their professional organizations unprepared for the demands of the changing medical marketplace (Malatesta 2011, p. 65). Experiences with mutual societies trying to secure medical services at reduced prices had been limited to a small number of fringe doctors in large industrial towns. Medical trade unionism had been limited in the main to *medici condotti* who had municipal authorities as their counterparts in negotiations about structure of practice, stability, and pension's right. Bruno Biagi, President of INFPS, described this transition as the "vanishing of the characteristics of the liberal profession, while doctors' duties as collaborators of the state on the highest function to secure the health of stock and the strength of the race keep increasing" (Biagi 1935). For private practitioners and municipal doctors in the process of making "*medici della mutua*" (doctors of the mutual fund) "the fight .... against insurance agencies" was perceived as "fateful and inescapable" (Pisenti 1930). Doctors demanded pay for services (*a notula*) instead of lump sums (*forfeit*) per patient or episode of illness and open lists of doctors instead of closed panels allegedly to allow patients' freedom of choice. Above all, doctors' fundamental request was "to limit insurance coverage to the economically needy" to let them cater to middle-class patients and not just for the well-to-do, who were the traditional preserve of the medical elite (Pisenti 1930). Moreover, doctors grew unhappy with the agreements stipulated with *casse mutue* and suspicious that their syndicate colluded with *casse mutue* on the lowest possible fee to ensure their financial stability. The tension peaked in 1937 when the president of the *Sindacato Fascista dei Medici* Enrico Morelli resigned over the contract which insured "at a miserable price" farm tenants and sharecroppers, the traditional self-paying clients of rural practitioners (Preti 1987, p. 288). Despite these important changes however, the organizational setting of medical practice changed little, and the solo practitioner remained the norm among the members of the profession.

Mussolini ostentatiously displayed a personal penchant for medicine. *Il Duce* frequently used medical metaphors in his speeches and defined governing "an art" and particularly "a medical art" (Ludwig 1932). Furthermore, he often posed himself as "the doctor of the Nation", who took care in "curing" both natural and political scourges ("socialist scrofula, giolittian plague, clerical scabies") and applied radical remedies with "the necessary cruelty, the cruelty of the surgeon" (Rigotti 1987). This however did not ensure Italian doctors the special position they enjoyed in Nazi Germany. Mussolini's speech to medical doctors at the National Congress of the Syndicate of Fascist physicians in 1931 helps in understanding how *Il Duce* conceived their role (Mussolini 1931). Mussolini insisted on the doctors' cultural power over ordinary people as a vehicle to spread fascist views over all matters of life, from food habits (consume rice, not wheat, to help the balance of payment) to women's standards of beauty (prefer plump to skinny, because they are more prolific) and even the duration of economic recession. Like priests, physicians were to be the agents cultivating the faith in Fascism, the *militi dell'Idea* (soldiers of the Idea-Gabbi 1931).

Of course, intensions do not always correspond with reality. Maria Malatesta has argued that doctors were the professionals least aligned with the regime, when compared, for example, with lawyers and engineers (Malatesta 2011, p. 68). However, as the regime consolidated its corporatist structure and its grip over society, *Ordine dei Medici* was dissolved, and all doctors enrolled in the fascist syndicate. For doctors in public offices party membership became mandatory and loyalty checks were required, as with the 1931 university professors' oath to the fascist regime (when only 14 resisted out of about 1200) and the 1938 declaration of non-Jewish origin as part of the infamous Racial Laws (Sarfatti 2018). Furthermore, restrictions of civil rights were imposed, and dissent was eradicated through confinement (internal exile), imprisonment, intimidation, violence, and at times murder. However, categories of resistance, collaboration, and coercion hardly capture the complex maze of relations of "mutual accommodations", which came to be negotiated between the various categories of doctors and scientists and the regime in different fields. Marla Stone has studied these complex relations of mutual exchanges and reciprocal legitimation in arts, architecture, and urbanism (see Stone 1998; see also Ben-Ghiat 2001 on a similar vein in cinema). In medicine, Mitchell Ash's dynamic framework of "self-mobilization" and "resources for each other" has proved useful in examining mutual advantages (Schmuhl 2011), including careerism and political opportunism, in the relations between professionals and specific policies (Weindling 1988). For example, certain professional categories were crucial for the implementation of policies of special import for the regime. This is, for example, the case of doctors who specialized in curing tuberculosis, or "*tisiologi*", who participated in the advantages of the rich sanatorial programme, and eugenists and demographers of various stripes who subscribed the *Manifesto della Razza* which paved the way to the Racial Laws. Psychiatrists and psychologists had previously forged privileged relations with the new men of the regime. The special relations carved out with nationalists and futurists during the First World War, and the prompt and enthusiastic following of the Rivoluzione Fascista of most of them (see, e.g., Ferrari 1922), granted them special influence and privileges. This included, for example, a massive project of large and luxurious *manicomi*, where they gained almost absolute power as "*gerarchi di manicomio*" (hierarchs of the asylum) (Giacanelli 2009, p. 82). By contrast, *medici condotti* had gained a subversive reputation in the early 1900s (Cogliolo 1899) and had followed the Socialist Party in opposing the war (Detti 1979). However, they also made up *the gran massa dell'armata medica* (the great mass of the medical army) and the closest to the public, and were therefore of great political import to the regime, as

Mussolini himself recognized in his address to the doctors (Mussolini 1931). Their tarnished reputation was first cleared by the fascist foundational myth of the trenches where most of them also fought, and their national organization was eventually subdued to the cause of Fascism through a long process of penetration (Soresina 1982). *Medici condotti* however maintained a lower social status compared with hospital doctors and academics who had a significant rise in social standing and political influence. They were granted considerable privileges in expanding private practice both within and outside public hospitals. Petragnani's law of 1938, which designed, for the first time, the organization of general hospitals long awaited since 1880, prohibited hospitals from limiting doctors' practice in private clinics and required hospitals to save 10 per cent of total beds for private patients. This regulation remained a structural feature of the hospital system in Italy up to the reform of dual practice in 1999.

Others championed some of the regime's policies, which met their professional capacities and career expectations, and were rewarded with special privileges and access to resources. This was particularly the case of doctors in scientific research, a traditional interest of the regime and of Mussolini himself (see, e.g., Turi 2016). Nicola Pende, for example, was a pioneer in the field of endocrinology and hormone therapy, widely recognized by the international community. His professional interests met with the fascist policy of race, which he championed in its "Latin" eugenic version (Turda and Gillette 2014). His theory of "orthogenesis" (Pende 1933) epitomized the optimal tool to ameliorate the race in a way that was both original in respect with the so-called Nordic, or Nazi, eugenics, which allowed to claim the superiority of the Italian science and was also accepted as fully compatible with Catholic values, which ensured its domestic acceptance (Cassata 2011, p. 138; Beccalossi 2020). This earned him a central role in fascist medicine, which was crowned by the founding of his *Istituto di biotipologia individuale ed ortogenesi umana*. The regime founded two of the most important Italian scientific institutions of the twentieth century, the *Consiglio Nazionale delle Ricerche* (CNR-National Research Council) and the *Istituto di Sanità Pubblica* (ISP-now Istituto Superiore di Sanità). The early years of their otherwise glorious history offer clear examples of self-mobilization of scientists and the development of exchanges of reciprocal legitimation between scientists' personal and/or professional interests and the regime's political goals. The early years of ISP and the role of its director Domenico Marotta in transforming a public health institution devoted to field research and training of medical personnel, as originally planned by the Rockefeller Foundation, its main sponsor, and Lewis Hackett, a long time collaborator in malarial control, into a centre of basic research at the intersection of politics, basic research, and the industry are examined in Cozzoli and Capocci (2011), while Stapleton analysed its relations with Lewis Hackett and the Rockefeller Foundation (Stapleton 2000). The CNR was founded to provide Italy with a large centre for applied research of national import, detached by the University, where substantial resources could be concentrated (Maiocchi 2003). However, despite the series of impressive presidents such as Guglielmo Marconi and general Pietro Badoglio, it barely survived until its internal management invented for the Institute a pivotal role in the politics of autarky of the regime and then in preparing war. This started a period of intense research in collaboration with national industry focused on textiles, agriculture, and alternative sources of energy to mobilize national resources for war. This research did not much help Italy in attaining self-sufficiency but put the Institute in the spotlight and energized its researchers. Maiocchi reports the enthusiasm of Giacomo Levi, who had studied for decades how to use Italian coal to replace imported anthracites, in finding that people were suddenly interested in what he had been doing from decades: "How different is the climate to-day! The indifferent is now enthusiastic, the incompetent rush to study to be a scholar, industrialists, technicians, capitalists are all mobilized" (Maiocchi 2009, p. 70). Research projects, which had been dragging for decades, were resumed by scholars thrilled by lavish funding and opportunities to be the focus of national policy, oblivious of its ultimate goals. These relations, which in many ways helped the regime to attain its goals, define what John Kriege, following Hannah Arendt, calls "the problem of evil" in the difficult task of re-developing on different grounds the European scientific community after the war (Kriege 2016). In Italy, it helps in understanding, on the contrary, the continuity in men and institutions in research and organization of health services between the fascist regime and the postwar democratic republic, which is the main theme of the next chapter.

References

Angelini, F. (1937) Gli sviluppi della mutualità sindacale Fascista. L'Assistenza Sociale, 11 (12), pp. 955–960

Barone, G. (1984) Statalismo e riformismo. L'Opera Nazionale Combattenti, 1917–1923. Studi Storici, 25, pp. 203–244

Beccalossi, C. (2020) Optimizing and normalizing the population through hormone therapies in Italian science, c. 1926–1950. British Journal of the History of Science, 53 (1), pp. 67–88

Ben-Ghiat, R. (2001) Fascist Modernities. Italy, 1922–1945. Berkeley, University of California Press

Bernhard, P. Klinkhammer L. Eds (2017) L'Uomo Nuovo del Fascismo. La costruzione di un progetto totalitario. Roma, Viella

Bevilacqua, P., Rossi Doria, M. (1984) Le bonifiche in Italia dal '700 ad oggi. Roma, Laterza

Biagi, B. (1935) Orientamenti del diritto corporativo nei rapporti della medicina e dell'igiene sociale. I Congresso della Previdenza Sociale. Bologna. Assistenza Fascista, 2, pp. 9–11

Bottai, G., Turati, A. (1929) La Carta del Lavoro illustrata e commentata. Roma, Ed. Diritto del Lavoro

Bryder, L. (1988) Below the magic mountain. A social history of tuberculosis in twentieth-century Britain. Oxford, Clarendon Press

Buccella, M.R. (1929) Lo svolgimento e il sistema della bonifica integrale. Giornale degli Economisti, 69, pp. 584–616

Caprotti, F. (2007) Destructive creation: fascist urban planning, architecture and New Towns in the Pontine Marshes. Journal of Historical Geography, 33, pp. 651–679

Caprotti, F. (2008) Internal colonization, hegemony and coercion. Investigating migration to Southern Lazio, Italy, in the 1930s. Geo Forum, 39, pp. 942–957

Cassata, F. (2006) Molti, Sani e Forti. L'eugenetica in Italia. Torino, Bollati Boringhieri

Cassata, F. (2011) Building the New Man. Eugenics, racial science and genetics in twentieth-century Italy. CEUP Press

Chang, N.V. (2015) The crisis-woman. Body politics and the modern woman in Fascist Italy. Toronto, University of Toronto Press

Cherubini, A. (1974) Storia e problemi dei rapporti fra tubercolosi e mutualità. Problemi della Sicurezza Sociale, 29 (2), pp. 167–201

Cogliolo, P. (1899) La questione dei medici condotti. Nuova Antologia, 83, pp. 508–516

Cohen, J.S. (1979) Fascism and agriculture in Italy. Politics and consequences. Economic History Review, 32 (1), 70–87

Condrau, F. (2010) Beyond the total institution. Towards a reinterpretation of the tuberculosis sanatorium. In: Tuberculosis then and now. Condrau, F. Worboys, M. Eds. Montreal & Kingston, McGill-Queen's University Press, pp. 72–99

Cozzoli, D., Capocci, M. (2011) Making biomedicine in twentieth-century Italy: Domenico Marotta (1886–1974) and the Italian Higher Institute of Health. British Journal for the History of Science, 44 (4), pp. 549–574

D'Antone, L. (1979) Politica e cultura agraria. Arrigo Serpieri. Studi Storici, 20, pp. 609–642
De Felice, R. (1974) Mussolini il duce. Gli anni del consenso, 1929–1936. Torino, Einaudi
Del Curto, D. (2010) La costruzione della rete sanatoriale italiana. In: Il Villaggio Morelli, Bonesio L. Del Curto D. Eds. Reggio Emilia, Diabasis, pp. 189–224
Detti, T. (1979) Medicina, democrazia, e socialismo in Italia fra '800 e '900. Movimento Operaio e Socialista, 2 (1), pp. 3–49
Dogliani, P. (1999) L'Italia Fascista, 1922–1940. Milano, Sansoni
Evans, H. (1989) European malaria policy in the 1920s and 1930s. The epidemiology of minutiae. Isis, 80, pp. 40–59
Eylers, E. (2014) Planning the Nation. The sanatorium movement in Germany. Journal of Architecture, 19, pp. 667–692
Falasca-Zamponi, S. (1997) Fascist Spectacle. The aesthetics of power in Mussolini's Italy. Berkeley, University of California Press
Ferrari, G.C. (1922) La psicologia della Rivoluzione Fascista. Rivista di Psicologia, 18, pp. 145–160
Fiorilli, O. (2016) Biopolitica dell'igiene nel primo dopoguerra. Genere e governo dei corpi nella costruzione della assistenza sanitaria visitatrice, Italia Contemporanea, 28 (2), pp. 209–232
Francioni, G. (1930a) Finalità e limiti dell'assistenza obbligatoria contro la tubercolosi in Italia. Difesa Sociale, 9, pp. 58–61
Francioni, G. (1930b) Il Dispensario antitubercolare e l'assicurazione obbligatoria. Difesa Sociale, 9, pp. 183–189
Francioni, G. (1957) Trent'anni di assicurazione obbligatoria contro la tubercolosi. Previdenza Sociale, 13, pp. 1059–1094
Fumian, C. (1983) I tecnici fra agricoltura e stato, 1930–1950. Melanges de l'Ecole Francaise de Rome, 95, pp. 209–217
Gabbi, U. (1931) Medicina politica e dottrina fascista. Politica Fascista, 3, pp. 444–458
Gachelin, G. (2013) The interaction of scientific evidence in debates about preventing malaria in 1925. Journal of the Royal Society of Medicine, 106 (10), pp. 415–420
Gachelin, G., Garner, P., Ferroni, E., Verhave, J.P., Opinel, A. (2018) Evidence and strategies for malaria prevention and control:historical analysis. Malaria Journal, 17 (1)
Garvin, D. (2015) Taylorist breast-feeding in rationalist clinics. Constructing industrial motherhood in Fascist Italy. Critical Inquiry, 41, pp. 655–674
Ghirardo, D. (1989) Building new communities. New Deal America and Fascist Italy. Princeton, Princeton University Press
Giacanelli, F. (2009) Gli psichiatri e il regime. Ipotesi per una ricerca. Rivista Sperimentale di Freniatria, 133, pp. 78–85

Gini, C. (1912) I fattori demografici nella evoluzione delle Nazioni. Torino, F.lli Bocca

Giorgi, C. (2004) La previdenza del regime. Storia dell'INPS durante il fascismo. Bologna, Il Mulino

Giorgi, C. (2012) The allure of the welfare state. In: G. Albanese, R. Pergher, Eds. In the society of Fascists. London, Palgrave Macmillan, pp. 131–148

Giorgi, C., Pavan, I. (2021) Storia dello Stato sociale in Italia. Bologna, Il Mulino

Gorsky, M. (2011) Local government health services in interwar England. Problems of quantification and interpretation. Bulletin of the History Medicine, 85, pp. 384–412

de Grazia, N. (1992) How Fascism ruled women. Italy, 1922–1945. Berkeley, University of California Press

Gualdi, A., D'Argenio, A. (1940) Sulla sorte dei tubercolotici dimessi dai sanatori. Difesa Sociale, 19, pp. 158–166

Hackett, L. (1925) The importance and uses of Paris Green (copper aceto arsenicate) as an Anopheles larvicide. Proceeds of the First International Congress on Malaria, October 4–6, Rome, pp. 1–15

Hackett, L. (1937) Malaria in Europe. An ecological study. Oxford, Oxford University Press

Ipsen, C. (1996) Dictating demography. The problem of population in Fascist Italy. Cambridge, Cambridge University Press

Istituto Centrale di Statistica del Regno d'Italia (1934) Statistica degli ospedali e degli altri istituti pubblici e private di assistenza sanitaria ospedaliera nell'anno 1932-XI. Roma, Istituto Poligrafico dello Stato

Koch, R. (1905) The current state of the struggle against tuberculosis. Nobel Lectures, Physiology or Medicine, 1901–1921. Amsterdam, Elsevier, 1967, pp. 169–178

Kriege, J. (2016) The "problem of evil" and postwar scientific cooperation in Europe. In: Defrance, C. Kwaschik, A. Eds La guerre froide et l'internationalization des sciences. Acteurs, reseaux et institutions. Paris, CNRS Editions, pp. 31–49

League of Nations (1939) European conference. Land Reclamation and improvement in Europe. Official N° 21 M 13

League of Nations. Malaria Commission. (1925) Report on its tour of investigation in certain European countries in 1924. Report CH273. Geneva, League of Nations

Ludwig, E. (1932) Colloqui con Mussolini. Verona, Mondadori

Luzzatto, S. (1999) The political culture of Fascist Italy. Contemporary European History, 8, pp. 317–334

Maiocchi, R. (2003) Gli scienziati del Duce. Il ruolo dei ricercatori e del CNR nella politica autarchica del Fascismo. Roma, Carocci

Maiocchi, R. (2009) Fascist autarky and the Italian scientists. Journal of History of Science and Technology, 3, pp. 62–73

Malatesta, M. (2011) Professional men, professional women. The European professions from the 19th century until today. London, Sage
Mantovani, C. (2004) Rigenerare la società. L'eugenetica in Italia dalle origini ottocentesche agli anni Trenta. Soveria Mannelli, Rubbettino
Melis, G. (1996) Storia dell'Amministrazione italiana, 1861–1993. Bologna, Il Mulino
Morelli, E. (1943) Presentazione. In L'Eltore, G. La mortalità tubercolare a Roma. La lotta contra la Tubercolosi 5, pp. 43–46
Murard, L., Zylberman, P. (1987) La mission Rockefeller en France et la création du Comité national de défense contre la tubercolose, 1917–1923. Revue d'historie modern et contemporaine, 34 (2), pp. 257–281
Mussolini, B. (1925) Lotta contro la malaria. In: Opera Omnia di Benito Mussolini, vol. XXII, Susmel E., Susmel D., Firenze, La Fenice, 1957, pp. 36–39
Mussolini, B. (1927) Il discorso dell'Ascensione. In: Opera Omnia di Benito Mussolini, vol. XXII, Susmel E., Susmel D., Firenze, La Fenice, 1957, pp. 367–390
Mussolini, B. (1928) Prefazione. In: R. Korherr, Regresso delle Nascite; Morte dei Popoli. Roma, Libreria del Littorio
Mussolini, B. (1931) Il discorso ai medici. In: Opera Omnia di Benito Mussolini, vol. XXII, Susmel E., Susmel D., Firenze, La Fenice, 1957, pp. 58–63
Nelson, E.J. (1991) Nothing ever goes well enough: Mussolini and the rhetoric of perpetual struggle. Communication Studies, 42 (1), pp. 22–42
Paxton, R.O. (1998) The five stages of Fascism. Journal of Modern History, 70, pp. 1–23
Paxton, R.O. (2004) The anatomy of Fascism. New York, Knopf
Pende, N. (1933) Bonifica umana razionale e biotipologia politica. Bologna, Cappelli
Pick, D. (1989) Faces of degeneration. A European disorder, c. 1848–c. 1918. Cambridge, Cambridge University Press
Pisenti, G. (1930) Lotte fra medici ed istituti assicuratori nell'assicurazione malattie. Storia di oggi, di ieri e di domani. Rassegna della Previdenza sociale, 17 (8), pp. 5–21
PNF-Partito Nazionale Fascista (1936) La Politica sociale del Fascismo. Roma, La Libreria dello Stato
Preti, D. (1987) La modernizzazione corporativa, 1922–1940. Economia, salute pubblica, istituzioni e professioni sanitarie. Milano, Franco Angeli
Quine, M.S. (2002) Italy's social revolution. Charity and welfare from liberalism to fascism. Basingstoke, Palgrave
Rigotti, F. (1987) Il medico-chirurgo dello Stato nel linguaggio metaforico di Mussolini. Istituto Lombardo per la Storia del Movimento di Liberazione. Milano, Cordani, pp. 501–517
Roatta, G.B. (1938) i cinquant'anni del Dispensario, 1887–1937. Lotta contro la Tubercolosi, 9, pp. 1115–1126

Salvati, M. (2006) The long history of corporatism in Italy. A question of culture or economics? Contemporary European History, 15 (2), pp. 223–244

Saraiva, T. (2011) Costruire il fascismo. Autarchia e produzione di organismi standardizzati. In: Scienza e cultura nell'Italia Unita. Annali 26. Cassata, F. Pogliano, C. Eds Torino, Einaudi, pp. 203–239

Sarfatti, M. (2018) Gli Ebrei nell'Italia fascista. Vicende, identità, persecuzione. Torino, Einaudi

Schivelbush, T. W. (2006) New Deals. Reflections on Roosvelt's America, Mussolini's Italy, and Hitler's Germany, 1933–1939. New York, Metropolitan Books

Schmidt, C.T. (1937) Land reclamation in Fascist Italy. Political Science Quarterly, 52 (3), pp. 340–363

Schmuhl, H.W. (2011) "Resources for each other". The society of German neurologists and psychiatrists and the Nazi "health leadership". European Archive of Psychiatry and Clinical Neuroscience, 261, pp. 197–201

Schneider, W.H. (1991) Quality and quantity. The quest for a biological regeneration in twentieth-century France. Cambridge, Cambridge University Press.

Snowden, F. (2006) The conquest of malaria. Italy, 1900–1962. New Haven, Yale University Press

Soresina, M. (1982) Dall'Ordine al Sindacato. L'organizzazione professionale dei medici dal liberalismo al fascismo, 1910–1935. In: Cultura e società negli anni del fascismo. Milano, Cordani, pp. 181–208

Stapleton, D.H. (2000) Internationalism and Nationalism: the Rockefeller Foundation, public health and malaria in Italy, 1923–1951. Parassitologia, 42, pp. 127–134

Stone, M. (1998) The Patron State. Culture and politics in Fascist Italy. Princeton, Princeton University Press

Turda, M., Gillette, A. (2014) Latin Eugenics in comparative perspective. London, Bloomsbury Academics

Turi, G. (2016) Sorvegliare e punire. L'Accademia d'Italia, 1926–1944. Roma, Viella

Weindling, P. (1988) Fascism and Population in comparative European perspective. Population and Development Review, 14, pp. 102–121

Wilson, P. (2002) Peasant women and politics in Fascist Italy. The Massaie Rurali. New York, Routledge

CHAPTER 5

Postwar: Roads Not Taken

Italy emerged from the Second World War divided, destitute, and debased. Italians were "exhausted and discouraged" (Lanaro 1997, p. 35). Misery and social disarray surpassed material destruction. The population was in dire need of food and medicines. War's disruption caused the spread of infectious diseases and the resurgence of typhoid, pulmonary tuberculosis, and malaria. The alliance with the Nazis had cut Italy out of the new miracle drugs such as penicillin, which was transforming the practice of medicine, medical research, and the pharmaceutical industry. The welfare system that fascism had built was "an imposing building in ruins" (Cabibbo 1944). However, in Italy "sad to say, the Government apparently participates in the awe which strikes people who face a dreadful disaster" (Missiroli 1945, p. 4). New public policies inspired by the concept of social security and social rights for all were transforming both the role of the modern state and the expectations of its citizens. Parties and governments of liberated Italy opposed to the new concepts and towards reform proposals maintained a stubborn silence, in defiance of the high principles established in the Constitution they participated in writing. Scholars and bureaucrats as well as politicians shunned foreign examples and domestic proposals to meet the social crisis and remedy the shortcomings of the ineffectual patchwork of services inherited from the fascist regime. This chapter briefly describes the complex political and institutional events of the transitional period from the fall of Fascism in 1943 to the establishment of a democratic republic in 1948. It then considers the war's impact

F. Taroni, *Health and Healthcare Policy in Italy since 1861*,
https://doi.org/10.1007/978-3-030-88731-5_5

on population health focusing on the project to control malaria in Sardinia and the aborted plan to develop a state-owned penicillin production plant. Finally, it examines the main welfare reform proposals, considering the reasons for the institutional "awe" that prevented much-needed change and ultimately led to a politics of strong continuity with the men and the institutions of the fascist regime, rather than looking towards international examples and domestic projects.

A Country in Ruins

Italy entered the war in 1940 and from 1940 to 1943 mainly fought a "parallel war" with its Nazi Ally invading the Balkans and Crete (Greece), but also participating in the disastrous Russian campaign. There was almost a decade of wars, from the invasion of Ethiopia in 1935 and the massive participation in the Spanish Civil War, to the invasion of Albania in 1939. War reached Italian soil in 1943, with the Allied invasion of Sicily, which prompted the end of the regime. A few key turning points mark the transition from the fall of Mussolini and the collapse of the Fascist Regime on July 25, 1943, to the establishment of the democratic Republic in 1948, when the new Republican Constitution came into force and the first free political elections were held. By mid-September 1943, the state had collapsed and the country was split into two. South of Naples was the Kingdom of the South under the protection of the Allied Military Government. The German Army occupied the central and northern parts of the country under the puppet *Repubblica Sociale Italiana* (Italian Social Republic). Anti-fascist parties had also formed in Rome, the *Comitato di Liberazione Nazionale* (CNL-National Committee of Liberation), which invested special powers in its northern branch *CLN-Alta Italia* (CLN-AI National Committee of Liberation of Upper Italy), the institutional offspring of the armed Resistance under German occupation.

The Referendum of June 2, 1947, concerning the choice of a Republic or a Monarchy showed that Italy was also politically split. The Republic garnered only about two million more votes than the Monarchy, which obtained a majority of 62 per cent in the south. Moreover, Italy was a country intensely polarized into two competing ideological factions, with the largest and most influential Communist Party of the western European countries, which had a real possibility of coming to power through the ballot box. The potential conflict with Christian Democrats (DC) and right-wing parties was initially contained by cooperation in governments of "national unity". As tensions between the United States and Soviet Union broke out in open confrontation, collaboration ended. In March 1947 US President Harry Truman announced his plan for the containment of the Soviet menace (the so-called Truman doctrine), and in May Socialist and Communist parties were excluded from unitary government. The first free political elections of April 18, 1948, were a turning point in Italian political history, the first "critical juncture" of republican Italy (Fabbrini 2008). International interest focused on Italian elections as a watershed of the early Cold War between the East and the West, and the United States "took off the gloves" (Miller 1983) actively supporting anti-communist forces. Christian Democrats received 48.5 per cent of the votes and won an absolute majority of 305 seats in the Chamber of Deputies out of 574 seats, establishing a dominant position in Italian politics.

Destruction of houses, roads, bridges, and rail tracks occurred when military combat concentrated on Italian soil with bombing campaigns, the advance of the Allies Army from the south, and the armed partisan resistance in the northern part under Mussolini's *Repubblica Sociale Italiana*. Bombardments left hundreds of thousands of Italian homeless; foodstuffs were in short supply, and the cost of living had multiplied more than twenty times from prewar levels. Productive capacity was about two-thirds of prewar levels in agriculture and 70 per cent in industry. During the brutally cold winter of 1946–1947 transport was paralysed, coal was in short supply, and the recovering industrial production slumped again. Agricultural yield fell in the subsequent hot and dry summer, and Italians had, for the third consecutive year, the lowest average food levels of all the western European populations (Judt 2005, p. 86). Hyperinflation spiralled prices upwards, and the cost of living peaked at forty-four times its prewar levels (Silei 2004, p. 35). Inflation and the devaluation of the currency impaired the fragile finance of *Casse Mutue* and depleted the reserves of welfare-based *Enti* at the same time that high unemployment curtailed their revenues from employers' and employee's contributions. Difficulties in hospital and healthcare financing and devaluation of cash benefits, including pension and disease allowances, further increased the problems in meeting the social and health needs of the Italians. Moreover, hospitals suffered "first by destruction and secondly by the systematic looting of beds, equipment, X-ray facilities, and ambulances by the retreating enemy" according to UNRRA's reports (cit. in Snowden 2006, p. 183). The United Nations Relief and Rehabilitation Administration (UNRRA), which provided emergency relief to prevent starvation and mass epidemics, effectively took on housing, feeding, clothing, and caring for the battered population. UNRRA's aid to Italy reached its peak in 1946 with a grant of 450 million US dollars covering, in addition to food, fuel, and agricultural and industrial materials and machinery, the whole cost of imported medical supplies (Ginsborg 1990, p. 472 note 21). Between 1947 and 1952, the European Recovery Program was the main instrument of Italian economic recovery and its harmonization with American interests.

Health and the War

At Christmas 1944 Myron C. Taylor, chairperson of American Relief to Italy, described the deprivation in food and medicine of liberated Italy in dramatic tones: "should an epidemic break out here, it would sweep all before it so low is the resistance of the people. Medicines are very scarce until recently practically non-existent" (quoted in Bud 2007, p. 85). The overall mortality rate peaked in 1944 at 15.9 per 1000 population compared to 13.4 in 1939 and then slowly declined to return to prewar levels by the end of the decade (Fig. 5.1) (Istat 1949; Tizzano, 1950). Excess mortality was due to infectious diseases, particularly pulmonary tuberculosis, which increased by 28 per cent, while malaria and typhoid increased by 54 and 44 per cent respectively and remained higher than in the prewar years until 1947 (Tizzano 1950). The increase in mortality for pulmonary tuberculosis interrupted the downward trend observed beginning at the turn of the century. Compared to 761 deaths per million population in 1939, in 1945 there were about 8000 excess deaths, a rate of 930 per million. Mortality rates for pulmonary tuberculosis quickly declined to 475 per million in 1948 and by an additional 70 per cent by 1953, without a parallel reduction in the number of cases (Drolet and Lowell 1955). As with the First World War, the malaria upsurge followed environmental disruptions, including damage to the drainage systems that kept mosquitoes in check and the repatriation of soldiers from malarial areas in Greece and Albania. In the Tiber Valley and the Pontine Marshes, the retreating Nazi Army destroyed the drainage pumps to flood the area and slow down the advance of the Allied Army (Snowden 2008). The reclaimed land was returned to nineteenth-century conditions, which promptly brought a major resurgence of malaria (Missiroli 1944).

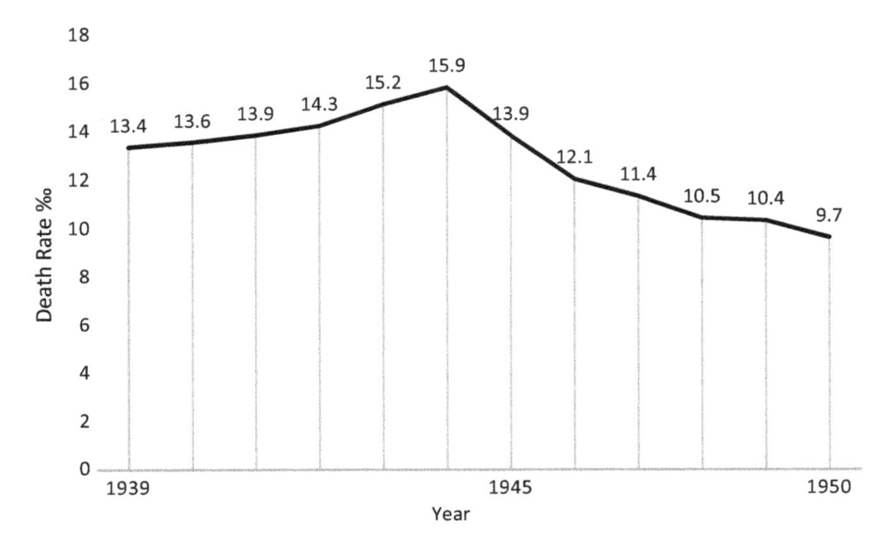

Fig. 5.1 Postwar: crude mortality rates per 1000 population, 1939–1950. (*Sources*: ISTAT, 1949; Tizzano, 1950)

The alliance with the Nazis had shut Italy out of the healthcare innovations brought about by the war, and many of the older drugs, sera, and vaccines were in short supply. The UNRRA programme of medical supplies provided insulin, diphtheria antitoxin, and toxoid, and newer drugs and biologicals, including penicillin and DDT, as well as information about their effective use (UNRRA 1946). The Tiber Delta and Castel Volturno, near Naples, were the first places where the residual spraying of houses with DDT to kill adult mosquitoes and the distribution of DDT on water surfaces from hand pumps and airplanes to kill the larvae was attempted (Soper et al. 1947). As DDT was found effective, easy to use, and inexpensive, the Allied Malaria Control Commission, under the direction of Paul Russell and Fred Soper of the International Health Division of the Rockefeller Foundation adopted this strategy in a large-scale programme to eradicate malaria by eliminating mosquitoes and their breeding places. Alberto Missiroli, the leading Italian malariologist, convinced Italian authorities to initiate the programme in Sardinia, the most malaria-ridden region with a high prevalence of *P. falciparum* (Tognotti 1996). The Sardinia Project became the showcase of a DDT-based malaria eradication programme around the world (Logan 1953). However, as organizational and political difficulties emerged (Tognotti 1995) and the first director John Kerr became convinced that eradication was not feasible, the ambitious goal of eradicating mosquitoes was changed into freeing Sardinia from malaria (Brown 1998). At the end of the project, for the first time in history, Sardinia had no reported new cases. However, breeding places remained, although drastically reduced. The strategy was extended to other parts of Italy, and succeeded in eradicating malaria, with the last reported endogenous case in 1962 (Snowden 2006, p. 210). This was the third anti-malarial strategy Italy had adopted in its history, from quininization in the early 1900s to *bonifica integrale* in the 1930s and DDT in the 1950s. At the international level, the outcome of the Sardinia Project of freeing Sardinia from malaria helped maintain the Rockefeller Foundation's concept of eradication. The concept prevailed in the eighth World Health Assembly of 1955, which voted to adopt DDT as the primary tool of the WHO Malaria Eradication Program, which however was largely unsuccessful (Packard 1997).

Penicillin was the quintessential "magic bullet", the product of the massive research and industrial effort of the war-years, the "good sister" of the bomb, and the fruit of "the Manhattan project for peace" (Lenoir and Hays 2000). Penicillin played a distinctive role in the postwar reconstruction of scientific research and the pharmaceutical industry in Italy (Taroni 2014). Initially, penicillin followed the American Army. By mid-1944, the journal *Il Policlinico* reported its first civilian use in Rome warning that "because of the high cost and short supply, it is safe to predict that civilians, particularly in Europe, will not have access to it until the end of the war" (Iandolo 1944). In fact, the US War Production Board released penicillin for commercial distribution to the public in March 1945 (Neushul 1993). A small note in the July 1945 issue of the same journal informed that UNRRA had allotted Italy 2500 vials of penicillin a month for the next five months, with indications limited to five severe clinical conditions (sepsis, meningitis, osteomyelitis, exposed fractures, and wounds) (UNRRA 1946). UNRRA also offered Italy a small pilot penicillin plant on February 1946, a combination of humanitarian relief and early cold war politics, as it was also offered to other countries similarly wavering between the opposing camps of the East and the West, including Poland, Czechoslovakia, Belarus, and Ukraine. Production was for domestic consumption, and export and commercialization in the private market were forbidden. The process of selecting the proper institution and the appropriate site was chaotic. The inauguration of the alleged *fabbrica italiana della penicillina* (penicillin Italian state plant) in Rome at the *Istituto Superiore di Sanità* took place almost two years later, on February 12, 1948, a few weeks before the fateful April elections. By then, however the situation had radically changed. Two private, commercial companies had been licensed to produce penicillin in Italy (Cozzoli 2014). Most importantly, Ernst B. Chain, one of the Nobel laureates from the original Oxford team who discovered penicillin, had joined the *Istituto Superiore di Sanità* (Cozzoli and Capocci 2011). The resources initially deployed for satisfying the needs of the Italian population were diverted into a world-class penicillin research centre. Limited production started in 1952, too late to help with the crisis caused by the American embargo on penicillin (and streptomycin) after the outbreak of the Korean War. Import of penicillin, and later streptomycin, was controlled by the High Commissariat of Public Health and managed by ENDIMEA, a state agency for public procurement financed in part by European Recovery Plan (ERP) counterpart funds.

A Lukewarm Transition

During the war and the early postwar years, the shortcomings of the existing system of social protection exposed by the crisis, along with the new concepts and examples of foreign models and the political climate of anti-fascist unity, seemed to provide the impulse for significant change in the welfare system inherited from fascism. Fascist social insurance schemes were in sharp contrast with the concept of social security and the universalistic and comprehensive principles of the multilateral organizations of the new world order, such as the Atlantic Charter, the Declaration of the Human Rights, and the WHO Constitution (Cueto et al. 2019, p. 34). Foreign examples of detailed plans for reform included primarily the celebrated Beveridge Report, issued in December 1942. Britain's Labour Party adopted this plan as its government programme in the summer of 1945. In the United States, President Harry S. Truman's special message to Congress proposed "a national health programme to provide adequate medical care for all Americans" (Blumenthal and Morone 2009, pp. 56–98). In France, Robert Debré for the Medical Committee of the Resistance envisaged a national health service based on a system of health centres that overhauled the hospital system and advocated the reform of medical education and the development of medical research (Immergut 1992, pp. 80–109).

In Italy, several proposals of various scope and detail emerged from different groups. A broad and heterogeneous movement of hygienists campaigned for the institution of a "technical" Ministry of Health (Del

Vecchio 1945). This resumed the traditional doctors' ambition for professional autonomy from political and administrative influences, which had been proposed in the aftermath of the First World War and instituted in Britain and France in 1919 and 1920 respectively. Hopes of reform were disappointed by the establishment of the *Alto Commissariato per l'Igiene e la Sanità Pubblica* (ACIS-High Commissioner for Hygiene and Public Health). The proposition submitted to the Constituent Assembly by the Fifth Sub-Commission "Health Organization" to include in the Constitution the institution of "a single technical administration distinct from other organs of the executive power" (Ministero della Costituente 1946) made ACIS appear a temporary setback, but a Ministry of Health was instituted only in 1958.

More ambitious in scope and depth was the plan approved in September 1945, a few months after the Liberation of Italy, by the CLN-AI—National Liberation Committee Upper Italy, which ruled the armed resistance against Nazi Germany and the fascist Republic in the North of Italy (Ministry of Health 1977). The plan was developed under the leadership of Augusto Giovanardi, a hygienist from the University of Milan, and envisaged a unitary national health system led by a Ministry of Health and Assistance, coordinated at the regional level and operated by municipalities. Regionalization would supplant *provincie* (the organizational centrepiece of the old system) as only the regional level could guarantee "the authority and expertise, from an administrative and technical point of view, necessary and sufficient to implement de-centralization". The regions would however only act as "communicating vessels, coordinated in their functions by a central body harmonizing the interests of each region with the general interest of the country". For such a goal, "a Directorate General under the Ministry of the Interior was no longer sufficient", but an "agile" Ministry of Health and Assistance with "coordinating and propulsive functions" was needed. After the long demise under fascism, municipalities were reinstated as the building blocks of the health system through *Uffici Comunali di Sanità* (Municipal Offices of Health). To make professional and technical autonomy compatible with social priorities and financial constraints, municipal health offices would submit to "political and administrative bodies" their budget with "expenditure allocated to their various health and welfare activities".

The remarkably "modern" design of a unitary and universal system functionally regionalized and locally operated neatly separated health services from the administration of *previdenza* consisting of social services financed by welfare-based *Enti*, such as INPS. However, by limiting its scope to services "directly financed by the state", consisting mainly of public health and ONMI's maternal and child services, it failed to integrate in the design ambulatory and hospital care financed and administered by *casse mutue*, which ran most of the health system. Beyond its content, however, the value of the CLN-AI plan was to channel the Resistance's ambitions into a political project for national renewal. This intent was not widely shared, as many were instead keen to compromise in the name of an orderly transition, or just because they were accustomed to their position and status. The proposal met fierce oppositions particularly from medical professionals and experts from many fields.

Opposition from the medical camp emerged at the first Congress of Hygienists held in Florence in 1946 and is duly reported in its Atti (Atti 1947). Further radicalizing his previous plan, Giovanardi explicitly called for a "reasoned and methodical demolition, according to a well-conceived plan, of the current system" (Giovanardi 1947, p. 64). The starting point would be the separation of *previdenza* based on social insurance, from healthcare, to be financed by general taxation, as with the Beveridge Report in Great Britain and the Wagner-Murray-Dingell bill in the USA (p. 49). Municipal Offices (which would in the main result from inter-municipal consortia, given the large number of small and very small *comuni*) were "functionally complete organs" in charge of preventive, diagnostic, and therapeutic services for 30,000 to 50,000 people. They would be staffed by a team of "resident doctors" attending specialist outpatient clinics and a small hospital with medical, surgical, and obstetric wards, and radiological and laboratory services for preventive and clinical activities. In essence, Giovanardi advocated a primary-care-led health service operated from health centres by full-time salaried personnel. His scheme shared many analogies with the Report of the British Medical Planning Commission of 1942, which derived in part from the Dawson Report (Webster 2003).

As with the British Report, all doctors' categories unanimously opposed the plan. *Medici provinciali*, who had benefitted in power and status from the rise of *provincie*, opposed both "the institution in each region of a *ministrino* (small minister) of health" and the dissolution of "the Provincial Health Office [...] the backbone of the country's public health and hygiene". Practitioners who would be the "resident doctors" of Municipal Offices proclaimed "their actual position unshakable" (Del Vecchio in Atti 1947). Both ACIS' representatives declared their firm opposition to the plan. Saladino Cramarossa, a former health officer in Rome who had risen to ACIS' director general, criticized the self-sufficiency of municipal offices and denied the usefulness of regionalization as "too broad to be effective". In his conclusions, the High Commissioner for Health Gino Bergami simply expressed his grievances towards a proposal he defined "naive and disorienting" and repeated calls to gradualism in considering changes to the existing system.

Bureaucracies from the existing social insurances also resisted the ideas of universal and uniform protection and opposed to major departures from the existing corporatist-occupational social insurance model. Influential bureaucrats expressed jointly their disposition towards expanding the existing system and preserving its socio-professional stratification and particularism (Santoro et al. 1946). Augusto Francioni, head of INPS' medical services, rose against the "totalitarian thesis" and its implicit universalism, which entailed the risk of reducing the benefits to which the insured were already entitled, which ought to be higher than those for the poor assisted by charities or municipal legal charity (Francioni 1946). Savoini, the future general manager of INAM, the main mutual fund of republican Italy, argued that state management of the "grandiose welfare activities" of *casse mutue* risked creating "excessive bureaucratization and overwhelming mechanization of assistance" and would result in expropriating the employers, whose contributions had helped building their patrimony. For Santoro, president of the National Insurance Association (INA), expanding social security to all citizens was "an extreme idea" which was "neither necessary nor useful in practice" as "the well-to-do have the means to deal with any event and risk, and can protect themselves through voluntary insurance". "Financial realism" induced the experts from trade unions and the left to shun foreign models as "our country cannot afford today the luxuries of a Beveridge or a Wagner-Murray-Dingell plan" (Giua 1946). The convergence of ideological and economic arguments across the political spectrum explains the scant appreciation of the Beveridge Report in Italy and its limited impact on postwar development of social security.

The Beveridge Report in Italy

According to Beatrice Webb, the Beveridge Report on Social Insurance and Allied Services was "a bomb thrown into the political arena" (Harris 1977). Its recommendations for a comprehensive and universal security system to win the "five giants" of want, disease, ignorance, squalor, and idleness in exchange for a modest weekly contribution from all attracted widespread attention far beyond the United Kingdom (see, e.g., Beland et al. 2021). It also reached fascist and anti-fascist circles in Italy. The fascist regime featured an image of the Report copying Italian and German social legislation, whose programmes had allegedly anticipated, and even exceeded Beveridge's recommendations. In 1942, Augusto De Marsanich's editorial in his journal *Assistenza Sociale* (Social Assistance), the official organ of the Fascist Confederations of Social Workers, pitied the "goodwill of Professor Beveridge". His main argument was that "the ideas inspiring his plan not only have no hint of originality whatsoever, but they are all drawn, with no worry of plagiarism, from the social doctrine of the enemy, from the Fascist doctrine" (De Marsanich 1942, p. 602). Others noted however "the unification and autonomy of the health services, which have been put together into a distinct national organization" and were interested in social security as "a means of redistributing national income" and "an instrument of national cohesion" (Del Giudice 1943). These significant points were apparently missed by the anti-fascist parties whose social policies were reviewed in 1944 (Cabibbo 1944). All parties paid lip service to the principles of national solidarity and liberation from need. Yet, major parties were in favour of covering only workers and their families, although Christian Democrats (DC) foresaw also "a potential gradual extension" of social insurance coverage which would "start from the categories most in need and worthy of support". For the Liberal Party, benefits should be proportionate to contributions, while for the DC they were to be differentiated by social category and place of living. The parties of the Left were mainly interested in participation of workers' representatives in the Boards of Directors of the reformed bodies, a symbol of "moral recovery of the new management" and of "rapprochement with the working masses". In summary, Cabibbo concluded, "no one wants to do away with the current system" and "no one is considering copying the Beveridge plan" (p. 47). Exceptions were the small circle of followers of Giuseppe Dossetti (Duchini 1949) and Agostino Gemelli (1948).

Luigi Einaudi was an economist of international renown and read the Beveridge Report through a scholarly lens. In his position as Governor of the Bank of Italy, Vice President of the Council of Ministers and Minister of the Budget, and, finally, as President of the Republic, he was also the architect of postwar economic policy, and his writings greatly influenced the political arena. In his *Lezioni di Politica Sociale* (Lessons in Social Policy) first published in 1949 but written in the last year of war, Einaudi judged with condescension the Beveridge Report. The Report had attained "the status of a legend, one of those legends that suddenly appeals to emotions" but was just "a summary, coordinating and expanding existing institutions and widely accepted principles" (Einaudi 1964, p. 89), aimed at "putting some order into the unspeakable vagaries of which the British legislation is scattered" (p. 86). He particularly disagreed with the cornerstones of the Report, unemployment, and healthcare policies. The programme against unemployment was thought to introduce barriers to competition, discourage work search, and impose brakes to economic growth. The healthcare programme was turned down on the curious argument that "the fight against disease, including injuries for the duration of medical treatment, has typically no bearing on insurance" (p. 106). Moreover, public funding of healthcare would dry up private charity and, by interposing a third party between the doctor and his patient, would destroy their relation of trust and induce suspicion, concluding that "sickness insurance costs and does not pay, and creates hatred and antisocial feelings" (p. 105).

In the wake of the institution of the much-awaited Commission for the Reform of Social Security, the dominant opinion was described as one of "substantial identity of views, not only among scholars and technicians, but also across political parties and trade unions" (Giua 1947, p. 43). The Italian "consensus" over social security reform was that the principle of universal and uniform coverage was an unnecessary, extreme, and luxurious extravagance. This was quite the opposite from the alleged British bipartisan "consensus" born from shared wartime experiences which eventually led to the institution of the British National Health Service (NHS). Richard Titmuss' classic studies of the effects of war on public opinion and party politics argued that bombing and the evacuation of London children led to a greater sense of national community and solidarity, which inaugurated a period of bipartisan consensus in major areas of public policy, including the Beveridgian welfare state (Timmins 1950, 1957). Revisionist accounts have shown that postwar political consensus was greatly exaggerated (see Jefferys 1987, Lowe 1990 and for health and the NHS, Webster 1990). Titmuss' arguments however help in explaining the "negative" consensus in Italy. At the end of the war, Italy was a divided nation, split between Fascist and Anti- Fascist forces and an insurgent North and a "liberated" South. Moreover, the key date of April 18, 1948, saw the most divisive election campaign in the country history. The "*consenso previdenziale*" boasted by the political forces could only bear a negative consensus on what ought to be excluded from the political agenda. The rich debate developing in the Constituent Assembly at about the same time had instead significant, although only long term, implications for health and social policy.

The Constitutional Debate

The debate over the Italian Constitution had a considerable impact on Italy's transition to democracy as well as on modern Italian culture and politics (Forlenza 2019, p. 220). Although the product of compromise between the various cultures represented in the Constituent Assembly (Thomassen and Forlenza 2016; Forlenza 2019), the Constitution firmly rejected fascist ideology protecting citizens' rights and preventing any form of concentration of power. The articles of the first part of the Constitution containing its Fundamental Principles are of quite a radical nature in terms of social, labour, and civil rights. However, many issues regarding the future society were ambiguous and open ended (Cartabia 2020).This is particularly the case with the health and social fields that are primarily distinguished by two articles. Article 32 is consistent with the principles of the Universal Declaration of Human Rights and particularly to the WHO constitution of 1946, which envisaged "the highest attainable standard of health as a fundamental right of every human being". The first paragraph of article 32 of the Italian Constitution declares health "a fundamental right of the individual and a collective interest" and enshrines this right into the responsibility of the Republic, including the state and local authorities (regions, provincie, and municipalities). Accordingly, the 1948 Constitution promised the devolution of considerable legislative power to the regions in several areas, including "public beneficence and medical and hospital care" (art.117). Article 38 deals instead with social assistance against the traditional misfortunes of life, such as old age, disability, and unemployment, which were to be based on social insurance except for the poor, whose "assistance" relies on the state through special legislation. As Amintore Fanfani, a major author of the article and future Minister of Labour and Social Providence (*Previdenza Sociale*), made clear "Previdenza does not include, as in the British plan, all citizens, but only those who are also workers" (cited in Pavan 2015, p. 99).

While article 32's reference to health as a personal right is universalistic in principle, article 38 maintains the traditional distinction between citizens and workers and between social insurance for workers and assistance for the rest of citizens which was a mainstay of the fascist scheme. The "workerist" (*lavorista*) interpretation of article 38, where personal social rights descend from work, prevailed for many years and helped maintain the sickness insurance schemes managed by *casse mutue*. A landmark decision of *Corte di Cassazione* (at the time, the Supreme Court) in February

1948, a few weeks after the entry into force of the Constitution, contributed to its "freezing", including the demotion of the potential impact of articles 32 and 117. The decision distinguished between the rules which had prompt implementation (*norme precettizie*—executive norms) and those, defined *norme programmatiche* (planning norms), whose implementation required, instead, supporting legislation and were therefore subject to political oversight. As a result, more innovative schemes remained dormant for decades. Regional autonomy, for example, was introduced in the early 1970s and was instrumental in the institution of the *Servizio sanitario nazionale*, which also implemented article 32 of the Constitution in 1978 (see Chap. 7).

The Rationalist Illusion: D'Aragona Commission

The history of the Commission for the Reform of *Previdenza Sociale* spans from the provisional government of general Pietro Badoglio in 1944 to the eve of the momentous elections of April 18, 1948. After a long series of false starts, the Commission took office on July 4, 1947, chaired by Lodovico D'Aragona an ex-socialist and former minister, concluded its work in a haste less than a year later, in February 1948, and submitted its eighty-eight final *mozioni* (propositions) on April 4 to Prime Minister Alcide De Gasperi (its detailed history is in Cherubini 1977, pp. 551–565). It was a period of growing international and domestic political tensions, and Italy was in the midst of its most severe postwar economic crisis, which was being "cured" by the severe measures of the "Einaudi line" (Spagnolo 1996). De Gasperi had hailed the Commission as "the Constituent Assembly of the Social security", and in receiving its conclusions, he praised their "balance" noting that the Commission "has rejected a model of insurance and social security for all, and the plan covers workers, both independent and dependent, and their families. The Commission has given up imaginative plans for the feasible [...] the prudence of the Commission stands out particularly where it has not aimed at a pettily leveling system" (Notiziario INPS 1949). The praised recommendations were however quickly and quietly put aside after the elections of April 18. The epitaph to the postwar reformist efforts came from De Gasperi's answer to a parliamentary question on December 15, 1948. Asked about the implementation of the D'Aragona's proposals, the Prime Minister was elusive: " I can not say when the reform will be submitted, because—and here we must put our hand on our heart—we are facing huge expenses and, if we seriously want a reform, we must simplify, we must make it lighter, don't keep it too heavy, because otherwise the state would have to shoulder a large burden" (www.degasperi.net).

Were the Commission's recommendations "prudent", as De Gasperi had initially acknowledged, or should they be "lightened" to prevent them overbearing the public coffer, as he later declared? Chairman D'Aragona had clearly asked the crucial question about who the beneficiaries should be: "all Italian citizens, all workers, workers below an income threshold, or only dependent workers" (D'Aragona 1947). The Commission's first three statements envisaged that social insurance should cover all public and private dependent workers with no income limit, followed by autonomous workers and finally their families. Furthermore, financing would not draw upon taxation but should rest on employers' and employee's contributions. For healthcare, the Commission recommended a general sickness insurance and proposed the unification of all forms of insurance under a single body, a scheme that was also in the original design of the Labor Charter of 1927. A heated controversy over the Commission's allegedly universalistic claim and its huge economic impact was sparked by a paper authored by Coppini, Emanuelli, and Petrilli, influential members of the Christian Democratic party, two of whom would later be President of INAM, the main Italian sickness fund. These "technocrats of the transition" from fascism to the republic (Pavan 2015, p. 94) alleged that the Commission excluded from coverage only a few "capitalists, beggars, vagrants and prostitutes", who were estimated at about 200,000 people, or just 1 per cent of the Italian population. The second criticism, and the principal issue, was that economic constraints should have been set prior to the definition of the beneficiaries. The argument was that "the boundaries within which the building could be erected" must always come first as "principles are often questionable and can change after discussion, but economic means are facts which impose limits to whatever initiative" (Coppini et al. 1948, p. 368). This reflected the austere economic planning of the time, when the Italian government diverted most of the American recovery funds to infrastructural investment, contrary to the Economic Cooperation Administration's advice culminating with the controversial Country Report of 1949 (Spagnolo 1996).

Prime Minister De Gasperi' s "hand on heart" quietly dismissed the gradualist motions of the Commission of D'Aragona, whose terse conclusions were buried under the massive 190 articles of Fanfani's "secret" reform project which never reached Parliament floor (Pavan 2015; see also Cherubini 1977, p. 373). In 1949 the conclusions of the Social week of the Italian Catholics exposed the ideological roots of the argument: "a general system of social insurance, that is one aimed at all citizens, is suited to a society which accepts the principles of a collectivistic economy, rather than a society that has rejected such principles" (ICAS 1950). Amintore Fanfani, now Minister of Labour and Previdenza sociale, was adamant in opposing previdenza sociale to the much broader concept of social security: "Previdenza is only a part of social security, to preserve the christian virtue of prudence and protect charity" (cited in Pavan 2015, p. 101). By 1950, the economic impossibility of a comprehensive reform was an accepted fact (see, e.g., Anselmi 1950). Organized medicine had also clearly made its preferences for maintaining the status quo evident. A "health referendum" of the recently restored *Ordini dei Medici* organized by the journal *Hygiene and Public Health*, and by the Parliamentary Medical Group had shown that "only a few *Ordini* agree on providing health care for all citizens, subject to a special health tax. Most are in favor of free care for the poor and the provision of care under a mutual or insurance basis for manual and intellectual workers" (D'Amico, 1947). The obvious conclusion was that "a system of protection for the entire population is not feasible today in Italy; the system must therefore address only the citizens who are economically weak" (Del Vecchio, 1950).

The widespread consensus eased the way to the quiet restoration of the "mutual order" and ensured the continuity of the institutions created under the fascist regime, which dominated the health policy of the Republic for the next thirty years. State apparatus and their bureaucracies had remained virtually untouched, including the semi-independent special agencies that had mushroomed under the fascist regime. This also the "problem of evil", that is, the collaboration with persons who maintained the significant positions in science and management they had obtained under the regime, as the *epurazione* (purging) of those who had been entangled with the fascist regime had been, at best, "faltering and uncertain" (Cartabia 2020, p. 95; see also Judt 2005, p. 47). None of the Commissione D'Aragona's gradualist recommendations were acted upon. The "freezing" of the Constitution made its article 38 on social assistance the reference point that, in the years to come, helped maintain a wall between citizens and workers, as well as between insurance for workers and assistance for "indigent" citizens. By the early 1950s, when the country was in a secure path to economic growth, the reform of healthcare was shelved. Roberto Maccolini's prescient comment sounds like an epitaph to a period of great expectations in which ideas of renewal were forced to give way to the early Cold war, the inertia of the institutions, and the power of vested interests: "we are following a path which certainly is not what other great civilized countries are taking. We are ascribing authority to a past although we are aware of the many shortcomings of which this past is rich" (Maccolini 1948, p. 94).

References

Anselmi, A. (1950) Aspetti economici e sociali delle assicurazioni sociali in Italia. Previdenza Sociale, 6, pp. 860–864.

Atti del primo congresso degli Igienisti Italiani (1947) Notiziario dell'Amministrazione Sanitaria, 8 (3)

Beland, D. Marchildon, G.P., Mioni, M., Petersen, K. (2021) Translating social policy ideas.The Beveridge report, transnational diffusion, and post-war welfare state development in Canada, Denmark, and France. Social Policy & Administration https://doi.org/10.1111/spol12755

Blumenthal, D. Morone, J.A. (2009) The heart of power. Health and politics in the Oval Office. Berkeley, University of California Press

Brown, P.J. (1998) Failure-as-success: multiple meanings of eradication in the Rockefeller Foundation Sardinia project, 1946–1951. Parassitologia, 40 (1–2), pp. 117–130

Bud, R. (2007) Penicillin. Triumph and tragedy. Oxford, Oxford University Press

Cabibbo, E. (1944) I partiti politici e la previdenza sociale in Italia. Rivista Infortuni e Malattie Professionali, 31 (6), pp. 13–48

Cartabia, M. (2020) The Italian Constitution as a revolutionary agreement. Italian Journal of Public Law, 12 (1), pp. 85–109

Cherubini, A. (1977) Storia della previdenza sociale. Roma, Editori Riuniti

Coppini, M.A., Emanuelli, F., Petrilli, G. (1948) Il costo della riforma della previdenza sociale. Rivista degli Infortuni e delle Malattie Professionali, 35 (3–4), pp. 367–445

Cozzoli, D. (2014) Penicillin and the European response to post-war American hegemony. The case of Leo-penicillin. History and Technology, 30 (1–2) pp. 83–103

Cozzoli, D., Capocci, M. (2011) Making biomedicine in twentieth-century Italy: Domenico Marotta (1886–1974) and the Italian Higher Institute of Health. British Journal for the History of Science, 44 (4), pp. 549–574

Cramarossa, S. (1947) La ricostruzione sanitaria. Igiene e Sanità Pubblica, 3, pp. 266–272

Cueto, M., Brown, T.M., Fee, E. (2019) The World Health Organization. A history. Cambridge, Cambridge University Press

D'Amico, D. (1947) I risultati del referendum sanitario. Igiene e Sanità Pubblica, 3, pp. 245–264
D'Aragona, L. (1947) Orientamenti programmatici della riforma della previdenza sociale. Previdenza Sociale, 6, pp. 201–205
De Marsanich, A. (1942) Note di politica sociale. L'Assistenza Sociale, 16 (12), pp. 597–603
Del Giudice, R. (1943) Il piano Beveridge "dalla culla alla bara". Le Assicurazioni Sociali, 1, pp. 1–17
Del Vecchio, G. (1947) Salus Publica Suprema Lex. Igiene e Sanità Pubblica, 1, pp. 49–72
Del Vecchio, G. (1950) E le stelle stanno ancora a guardare…Note e commenti al II° Congresso Nazionale della Protezione Sociale (Roma, 27–30 aprile 1950). Igiene e Sanità Pubblica, 6, pp. 160–176
Drolet, G.J., Lowell, A. (1955) The first seven years of the antimicrobial era, 1947–1953. American Review of Tuberculosis, 72, pp. 419–452
Duchini, F. (1949) I Piani Beveridge e la lotta contro la miseria. Cronache Sociali, 2, pp. 9–11
Einaudi, L. (1964) Lezioni di politica sociale. Torino, Einaudi
Fabbrini, S. (2008) De Gasperi e la "giuntura critica" del periodo 1948–1953. L'Italia dell'immediato dopoguerra fra due modelli di democrazia. Ricerche di Storia Politica, 1, pp. 53–67
Forlenza, R. (2019) On the hedge of democracy. Italy 1943–1948. Oxford, Oxford University Press
Francioni, G. (1946) L'assicurazione obbligatoria nella lotta contro la tubercolosi in Italia. Notiziario della Amministrazione Sanitaria, 7 (2), pp. 7–14
Gemelli, A. (1948) La esemplare organizzazione dell'assistenza sociale in Inghilterra. Vita & Pensiero, 9, pp. 20–34
Ginsborg, P. (1990) A history of contemporary Italy. Society and Politics 1943–1988. London, Penguin Books
Giovanardi, A. (1947) Riforma dell'ordinamento sanitario. Notiziario dell'Amministrazione Sanitaria, 8 (3), pp. 41–70
Giua, S (1946) I nuovi orientamenti della previdenza sociale. Previdenza Sociale, 5, pp. 163–167
Giua, S. (1947) Premessa ai lavori per la riforma della previdenza sociale. Previdenza Sociale, 2, pp. 43–46
Harris, J. (1977) William Beveridge. A biography. Oxford, Oxford University Press
Iandolo, C. (1944) Un nuovo chemioterapico: la penicillina. Il Policlinico, 51(27–31), p. 427
ICAS—Istituto Cattolico di Attività Sociali (1950) Atti della XIII settimana sociale dei Cattolici Italiani. Roma
Immergut, E.M. (1992) Health politics. Interests and institutions in Western Europe. Cambridge, Cambridge University Press

Istat—Istituto Centrale di Statistica (1949) Le cause di morte in Italia nel decennio 1939–1948. Bollettino mensile di statistica, 24 (10-11-12)
Jefferys, K. (1987) British politics and social policy during the Second World War. Historical Journal, 30, pp. 123–144
Judt, T. (2005) Postwar: a history of Europe since 1945. New York, Penguin
Lanaro, S. (1997) Storia dell'Italia Repubblicana. Venezia, Marsilio
Lenoir, T., Hays, M. (2000) The Manhattan project for biomedicine. In: Sloan P.R. Ed. Controlling our destinies. Historical, philosophical and ethical perspectives on the Human Genome Project. Indiana, University of Notre Dame Press, pp. 29–62
Logan, J. (1953) The Sardinia Project. An experiment in the eradication of an indigenous malarious vector. Baltimore, John Hopkins
Lowe, R. (1990) The Second World War: consensus and the foundation of the British Welfare State. Twentieth Century British History, 1(2), pp. 152–182
Maccolini, R. (1948) Riforma sanitaria e prassi sindacale. Rivista Italiana di Igiene, 8, pp. 81–97
Miller, J.E. (1983) Taking off the gloves. The United States and the Italian elections of 1948. Diplomatic History, 7 (1), pp. 35–55
Ministero della Costituente (1946) Relazione definitiva della V Sottocommissione "Organizzazione Sanitaria". Notiziario dell'Amministrazione Sanitaria, 7 (1), pp. 171–193
Ministero della Sanità Centro Studi (1977) L'istituzione del Servizio sanitario nazionale. Roma
Missiroli, A. (1944) La malaria nel 1944 e misure profilattiche previste per il 1945. Rendiconti Istituto Superiore di Sanità, 7, pp. 616–641
Missiroli, A. (1945) La sanità e la previdenza sociale nel quadro della ricostruzione. Igiene e Sanità Pubblica, 1, pp. 3–6
Neushul, P. (1993) Science, government and the mass production of penicillin. Journal of the History of Medicine and Allied Sciences, 48, pp. 371–395
Notiziario INPS (1949) Dichiarazione del Presidente del Consiglio sulla riforma della Previdenza Sociale, La Rivista Italiana di Previdenza Sociale, 2 (1), p. 87
Packard, R.M. (1997) Malaria dreams: postwar visions of health and development in the Third World. Medical Anthropology, 17 (3), pp. 279–296
Pavan, I. (2015) Un progetto "clandestino" di riforma. Fanfani e la Previdenza sociale. Contemporanea, 18, pp. 91–113
Santoro, G., Nervi, G., Savoini, V. (1946) La riforma della Previdenza Sociale. Contributi, discussioni, proposte. Previdenza Sociale, 31 (2), pp. 52–62
Silei, G. (2004) Lo stato sociale in Italia. Storia e documenti. II. Dalla caduta del fascismo ad oggi, 1943–2004. Bari, Manduria
Snowden, F. (2006) The conquest of malaria: Italy, 1900–1962. New Haven, Yale University Press

Snowden, F. (2008) Latina Province, 1944–1950. Journal of Contemporary History, 43 (3), pp. 509–526

Soper, F.L., Knipe, F.W., Casini, G., Riehl, L.A., Rubino, A. (1947) Reducing of anopheles density effected by the preseason spraying of building interior with DDT in kerosene, at Castel Volturno, Italy in 1944–1945 and in the Tiber Delta in 1945. American Journal of Tropical Medicine and Hygiene, 27, pp. 177–200

Spagnolo, C. (1996) La polemica sul "Country Study" il fondo lire e la dimensione internazionale del Piano Marshall. Studi Storici, 37 (1), pp. 93–143

Taroni, F. (2014) The fabbrica della penicillina in postwar Italy. An institutionalist approach. Medicina nei Secoli, 26 (2), pp. 639–662

Thomassen, B., Forlenza, R. (2016) Catholic modernity and the Italian Constitution. History Workshop Journal, 81 (1), pp. 231–251

Titmuss, R.M. (1950) Problems of social policy. London, HMSO

Titmuss, R.M. (1957) War and social policy. In: Essays on the welfare state. London, Allen & Unwin, pp. 75–87

Tizzano, A. (1950) L'andamento della mortalità e delle cause di morte nel decennio 1939–1949. Annali della Sanità Pubblica, 5, pp. 1513–1548

Tognotti, E. (1995) Americani, comunisti e zanzare. Il piano di eradicazione della malaria in Sardegna tra scienza e politica negli anni della guerra fredda, 1946–1950. Sassari

Tognotti, E. (1996) La malaria in Sardegna: per una storia del paludismo nel mezzogiorno, 1880–1950. Milano, F. Angeli

U.N.R.R.A. Missione Italiana in collaborazione con l'Alto Commissariato per l'igiene e la Sanità Pubblica (1946) Istruzioni sull'uso dei medicinali forniti dall'U.N.R.R.A. Notiziario dell'Amministrazione sanitaria, 7 (1), pp. 140–170

Webster, C. (1990) Conflict and Consensus. Explaining the British health service. Twentieth Century British History, 1, pp. 115–151

Webster, C. (2003) Medicine and the welfare state, 1930–1970. In: Cooter, R., Pickstone, J., Eds. Companion to medicine in the twentieth century. London, Routledge (pp. 125–140)

CHAPTER 6

The Rise and Fall of the Mutual Jungle

After the chaotic aftermath of the Second World War, Italy entered its golden age of economic growth, more rapid than any other European country and at any time in Italian history. This peaked in the "economic miracle" circa 1958–1963, after which an economic crisis interrupted fifteen years of sustained growth. Economic development brought about structural social changes and higher and new levels of personal consumption. The state had played an important part in stimulating the economy but failed to manage the social consequences of economic growth in housing, education, and public services. According to the *Economist*, with the economic miracle "Italy was dragged kicking and screaming into the middle of the twentieth century, from her agricultural structure to a full manufacturing organization". The "raging and chaotic" development of Italian society and the spectacular increase in private consumption had worsened traditional imbalances of the country. This prompted an "opening to the left" which brought to power coalition governments of the centre-left (*centro-sinistra*) with the participation of the Socialist Party, after a decade of centrist governments (Ginsborg 1990, p. 240). The "philosophical manifesto" (Lanaro 1997, p. 333) of the new government was the "Additional Note" (*Nota aggiuntiva*) to the General Report on the country's economic situation drawn up by the Budget Minister Ugo La Malfa in 1962. The Note adopted "the politics of planning" to strengthen the capacity of the state for "structural reforms" (as they came to be called—Ginsborg 1990, p. 265) prioritizing collective needs over

F. Taroni, *Health and Healthcare Policy in Italy since 1861*,
https://doi.org/10.1007/978-3-030-88731-5_6

individual consumption. Health was one of the main areas where "the direct action by the State" should focus, not only "to expand existing structures" but for "qualitative transformations of great importance" (Crainz 2003, p. 21). These grand purposes however did not help prevail over *casse mutue* and reform the health system.

Casse mutue (mutual health insurance funds, sickness funds) had mushroomed into a tangled "jungle" (Colombo 1977), a "fearful forest" (Cherubini 1977) of hundreds of agencies purchasing domiciliary and hospital services for their own insured population from thousands of contracted doctors and independent, mostly municipal, hospitals. Each sickness fund was a centre of power serving the interest of the factions of the coalitions in power. The Istituto Nazionale Assicurazione Malattia (INAM-National Institute of Sickness Insurance), the largest and most powerful of them, was "the strongest party in the country" and its President "the true Minister of Health" (Delogu 1967). Over the years, *casse mutue* expanded the breadth and depth of their coverage, reaching over 92 per cent of the Italian population. Always in the verge of financial collapse, economically inefficient, and totally disintegrated, *casse mutue* survived the explosion in drug expenditures of the 1950s but were overwhelmed by the cost of care in hospitals that were catching up with modern medicine, the expectations of their staff, and the demand of the population. Hospitals rapidly modernized in staff and organization under pressure from medical unionism and public opinion, and medical care "hospitalized". Massive increase in hospital expenditure brought down the precarious balances of *casse mutue*, which called for repeated state interventions. Their financial crisis opened the way to a long-awaited healthcare reform on an entirely new basis.

A New Old Policy

Out of the chaotic debate of the early postwar years, the continuity of the old state and its leaders prevailed over those who sought a renewal. The iconic representation of the continuity of the healthcare state was the passing of the law n.453 on May 13, 1947, which transformed *Ente Mutualità Fascista: Istituto Nazionale per l'assistenza di malattia* of 1943 into *Istituto Nazionale Assicurazione Malattie* (INAM—National Institute of Sickness Insurance). Following article 38 of the Constitution instead of its article 32, the logic of the health system privileged workers' rights to insure themselves against the risks to their labour capacity instead of citizens' rights to health protection. The rights of the worker-man were opposed to the needs of the citizen-man, always referred to in the masculine, according to the custom prevailing in the world of *casse mutue* where policyholders were overwhelmingly male. Casse mutue reflected the central characteristics of the welfare state that had matured during the fascist regime. Its Bismarckian structure implied that eligibility criteria, contributions, and levels of benefits varied by sector of employment, job, and rank. The male-breadwinner orientation assumed a family where the husband held a stable employment and the wife took care of the household.

Savoini, in his new position of director general of INAM, expressed the new old logic, impervious to Marshall's concept of the social rights of citizenship: "how to put at the same level, to entitle to the same benefits those who deny sociality by living as parasites without working and those who instead contribute to the common good with their work? There can be no such equality" (Savoini 1955, p. 40). This applied also to the level of benefits as each worker was entitled to maintain "the status he has achieved with his work". The main architect of the new policy was Giuseppe Petrilli, President of INAM from 1949 to 1958 and then Minister and European Commissioner. His strategy foresaw that "the expansion of *casse mutue* should proceed gradually, first testing the new legislation on groups numerically limited and professionally homogeneous" (Petrilli 1953, p. 253). The provision of sickness insurance had to be closely linked to jobs, and all workers employed in a particular industry were to be "automatically" insured by the same sickness fund. The principle of "automatic" enrolment also prevented competition among sickness funds. The new policy explicitly aimed at a two-tier system, consisting of "a minimum standard common to all" guaranteed by social insurance and complemented by voluntary "supplementary" insurance.

Coverage progressively expanded from 15.9 million people in 1943 to ever-larger sections of the population to reach 51.4 million people or 92.7 per cent of the Italian population in 1974, the year of dissolution of *casse mutue* (Table 6.1). The expansion resulted from increasing both the number of dependent workers and their families (usually limited to spouses and children) enrolled in the four main economic categories of agriculture, industry, commerce, and credit, and in the institution of new *casse mutue* to address the demands of special interest groups. An example of the latter is the extension of coverage to autonomous workers, which required state funding to substitute for employers' contribution. The constitution of the powerful new *casse mutue* of *Coltivatori diretti* (farmers cultivating their own land) in 1954, craftsmen (1956), and shopkeepers (1960) was part of the politics in favour of the *ceti medi* (middle class), which was a mainstay

Table 6.1 Population covered by *casse mutue*, 1943–1974, selected years

Year[a]	*Population covered*	
	Nr. (million)	*%*
1943	15.9	35.5
1950	17.8	37.9
1955	35.1	72.9
1960	40.1	80.0
1965	43.3	82.3
1970	48.0	88.5
1974	58.8	92.7

[a]or nearest year

Sources: Giorgi and Pavan (2021), Brenna (1974), V. Colombo (1977)

of the Christian Democratic party (Ginsborg 1990, p. 153). Moreover, dozens of usually very small categories of workers were covered as part of INAM with separate funding and management. This was the case, for example, for the occupational categories of domestic servants (created in 1952), fishermen of small-scale fishing in inland waters and their families (1958), and apprentices between fifteen and twenty years old (1955). Fragmentation limited solidarity across categories and resulted in an archipelago of autonomous organizations of various sizes, governed by specific laws and by-laws. The archipelago included highly centralized national bodies, decentralized confederations, at times on a territorial basis, as well as mammoth agencies organizing dozens of different categories, each with distinct funding and management as in the case of INAM.

In the early 1960s, 35 national agencies and 280 smaller *casse mutue* covered slightly less than 90 per cent of the Italian population (CNEL 1963). The National Institute of Health Insurance (INAM) was the first and most important health insurance agency in terms of the number of policyholders, financial capacity, and political influence. At the time of its dissolution in 1974, INAM enrolled over thirty million dependent workers or 55.8 per cent of the total population. Autonomous workers amounted to one quarter of the insured population and were organized by three *casse mutue* for farmers (five million), artisans, and shopkeepers (about four million each).

The issue was not only the extent of coverage, but also the variation in benefits and contributions. The principal factors in determining benefits and contributions were being the policyholder or a family member, the occupational category (autonomous or dependent workers of the traditional corporative grouping of agriculture, industry, trade, and banking and insurance services), and finally rank, where the main separation was between manual workers and clerks. Benefits included both compensation for lost income due to illness and medical care. Visits from general practitioners and hospital admissions were always covered but sickness allowances, drug prescriptions, and outpatient specialists differed from sickness fund to sickness fund, and even within them. A striking example of the disparity between economic sectors and occupational categories within the same agency is offered by INAM before its "little" reform of 1959 (Coppini 1959). All INAM policyholders were entitled to doctor's visits and hospital care, less than one-third were also entitled to sickness allowances, and three-quarters to pharmaceutical care, while integrative care such as prostheses, dental and optical care, spa, and so on, was entrusted to the discretion of Provincial Committees. Workers' contributions differed among sickness funds as a percentage of wages, the presence of ceilings, and method of calculation (by lump sum, by day or by week of work, or even acreage of land worked). In 1960, the average annual premium was 20,166 lire, ranging from 2123 to 42,022 lire for the agricultural and credit sector respectively, with over 28,000 lire for industrial workers and approximately 26,000 lire from pensioners. In times of crisis, the state granted various categories temporary exemptions and cuts in contributions, which were transferred to general taxation.

Each sickness fund contracted with its panel of doctors (both general practitioners and outpatient specialists) who were paid either a lump sum per patient or episode of illness or on a fee-for-service basis. Schedules of activities and prices varied by sickness fund. Sickness funds contracted with hospitals affiliated with their network. Hospital payment was on a per diem basis, with rates calculated according to a national formula excluding capital expenditures and depreciation. The cost-based system of payment was administratively simple but provided incentives for longer lengths of stay, which sickness funds tried to counter with periodic inspections.

Tuberculosis insurance continued to live a life of its own. It enjoyed great stability over time, protected as it was from the recurring financial crises that afflicted other *casse mutue* by the solid INPS' finances. It expanded to a limited degree from its prewar status to cover slightly less than twenty-eight million people, approximately 57 per cent of the Italian population (Papa 1958). Unlike sickness funds, tuberculosis insurance never crossed the frontier of self-employed workers and successfully resisted various attempt to merge with general insurance. Results were huge differences in monetary and service entitlements for patients with tuberculosis enrolled in sickness funds and those insured with INPS. While INPS policyholders with "active" tuberculosis were entitled to unlimited hospitalization and to economic indemnity which extended also to their spouses and children, persons with tuberculosis enrolled with INAM had a time limit of 180 days of hospitalization per calendar year and no economic subsidy. Moreover, fragmentation and differences in entitlements caused conflicts of jurisdiction between agencies which transformed the hospitalization of tuberculosis patients into occasions of "civil wars between agencies", according to the definition in vogue in the 1950s (Berlinguer and Delogu 1959).

The only measure which mitigated some of the most glaring inequalities in entitlements was the so-called *piccola riforma* (small reform) of INAM, significantly approved by its own Board of Directors in 1958, in the same year of the establishment of the Ministry of Health (Coppini 1959). Differences in entitlements and contributions between the various occupational categories covered by INAM dated back to the original 1943 law, which complied with the employment contracts then in force for each of the four fascist Corporations, deferring their harmonization to further regulation. However, the implementing regulation was never issued, and the republican law of 1947 limited itself to defer once again to further ordinary legislation. With a much-discussed initiative, INAM's Board of Directors significantly changed the internal regulations in terms of benefits and contributions. Among the many innovations was coverage for hospital birth delivery for uncomplicated or "physiological" childbirth, and the uniformation and extension of the "depth" of hospital coverage. A uniform time limit for hospital stay was set at 180 days a year for all insured persons and for all diseases, including "chronic diseases" and some "diseases specific of the old age" for pensioners.

The most significant innovation of the piccola riforma was the definition of a *Prontuario Terapeutico* (Therapeutic Formulary) listing the drugs to which those insured with INAM had access. This was a model for dozens of other national and local initiatives for the years to come regulating drug use and curbing pharmaceutical expenditures (Scarpa and Chiti 1975). The original idea was to "liberalize" doctors' drug prescription by eliminating prior authorization, select for entitlement only drugs with "a proven therapeutic value", and impose a co-payment on drug use. A commission classified branded drugs by broad therapeutic category, groups differentiating combinations from single action drugs, and subgroups of similar drugs with different prices. Ineffective drugs were to be excluded from coverage and drugs priced in excess of their subgroup average price subject to co-payment. The first example of "rational rationing" of drugs apparently backfired as the number of "medical specialties" or branded drugs covered by INAM rose from 7038 to 14,942, and only 3357 were subject to co-pay (INAM news 1959). The *piccola riforma* was however a major political initiative which qualified INAM as an "autarchic self-regulating body" (Tronelli 1973) well beyond what legislation allowed and gave it the lead of the entire sector, negotiating on equal footing with the government of the day. This was a clear message that the long-awaited institution of the Ministry of Health would hardly change the political landscape.

The Ministry of "salvo"

The compartmentalization of health interventions in the three parallel circuits of prevention, hospital and domiciliary medical care, and the fragmentation of *casse mutue* was clearly in need of direction and coordination. Such a function had been lacking since the time of the Constituent Assembly when ACIS was "temporarily" installed as an agency with the Presidency, but it took thirteen more years for the institution of a Ministry of Health. The 1949 Report of the International Commission of Experts convened in 1949 with the support of the Rockefeller Foundation suggested "a coordinating body, chaired by the President of the Council of Ministers and composed by the High Commissioner for Hygiene and Public Health and the Ministries directly or indirectly involved with activities in the field of hygiene and social assistance". A further report from the "Office for the reform of the public administration" took four years to reach the floor of Parliament and was then postponed for two more years for "coordinating its amendments" (Melis 1996, p. 437). The law was approved on March 13, 1958, and entered in force four months later.

The Ministry was given a token "supervisory" role in various fields of health and healthcare, which earned it the qualification of the Ministry of "salvo" (exceptions), as each definition of responsibility was quickly bounded by wide exceptions and strict limitations. It was entrusted with responsibilities in matters of hygiene and prevention of infectious diseases, but the financing of related activities, including hospitalization and vaccinations, remained in charge of the Municipalities. *Casse mutue,* including INAM itself, as the other welfare-based *Enti* INPS and INAIL, by virtue of their qualification as social agencies, answered to the Ministry of Labor and Social Security, which also maintained control over health and safety in the workplace. The Ministry of Health's "technical" supervision over hospitals excluded administrative and financial affairs, which remained firmly in the hands of the Ministry of the Interior, while building and renovation plans rested with the Ministry of Public Works. The marginal role reserved to the new Ministry was mirrored in the size of its budget, which amounted to just 43,882 billion lire, less than one-tenth of INAM's budget. Ambitions of "functional integration" with the goal of "harmonizing" entitlements and contributions failed miserably, including the attempt of the first Minister of Health, the Christian Democrat, and eminent pneumonologist Vincenzo Monaldi, to merge insurance against tuberculosis with general sickness insurance. Resistance always prevailed, as *casse mutue* were, in the words of the next Minister of Health, the socialist Luigi Mariotti, "centers of power that the State does not have the strength and the capability to control, because the mutual system is now a true political party, the strongest party in the country" (Mariotti quoted in Bruzzese 1967, p. 116).

The Dynamics of Spending and Service Provision

In 1974, health expenditures were estimated at 6134 billion lire, 6.4 per cent of the Gross National Product (Colombo 1977). *Casse mutue* accounted for 4534 billion or 74 per cent of the total. In 1952, the first year for which reliable data are available, *casse mutue* covered over 60 per cent of the population and accounted for 34.9 per cent of expenditures, compared to private expenditure estimated at 42.7 per cent and public expenditure at 22.4 per cent. Between 1952 and 1965, spending by the *casse mutue* quadrupled from 219 to 914 billion while public spending doubled, and private spending decreased by one-third, from 43 to 13 per cent of the total (Berlinguer 1968). In 1965, *casse mutue* covered about 45 million people, while the legal charity of municipalities and provinces still provided medical assistance to 2.6 million people, mainly in the south of the country (Somogyi et al. 1974). Between 1966 and 1974, health expenditure skyrocketed. Total spending increased at an average annual rate of 13.5 per cent; *casse mutue* spending grew even faster at an annual rate of 16 per cent, mostly driven by hospital expenditure, which rose from 831 to 1815 billion (Brenna 1974).

The time trends of *casse mutue* expenditures, including sickness allowances, doctors' fees, drugs, and hospital care, show a dynamic intertwining of changes in the number of insured persons and of their eligibility for various services, as well as of the volume of service use and their unit costs. To examine the profile of variation in sector-specific expenditures, it is appropriate to consider INAM data, which cover a large population and a long time span, and are broader in scope from other casse mutue. Between 1943 and 1949, coverage increased only slightly, from about twelve to fourteen million people, and sickness allowances were the main drivers of expenditures. Cash benefits accounted for 37 per cent of expenditures, compared to 22.5 per cent for hospital care and 13.7 per cent for doctors' fees. Expenditures for drugs were only 7.5 per cent of the total, reflecting the belated availability in Italy of the new branded drugs and the prescription primarily of relatively cheap *preparazioni galeniche* (non-prescription drugs). This profile reflects the most basic insurance function of income stabilization by protecting the wage earner against wage loss during sickness.

During the expansionary phase of the 1950s, the number of INAM policyholders increased by 60 per cent, from fourteen to twenty-one million, at a sustained rate of 10 per cent per year from 1949 to 1957. This was associated with the beginning of the economic miracle and the massive increase of occupation, particularly in industry, while agriculture suffered and lost part of its insured. Revenues of sickness funds increased from high employment and better wages, and coverage was expanded to spouses and children of the policyholders and, for the first time, to some categories of pensioners, opening insurance to a high-use, and high-cost segment of the population. This was also the period when an array of new, branded and expensive, heavily marketed drugs invaded the Italian market. Penicillin and streptomycin were quickly followed in the early 1950s by steroids, antipsychotics, oral anti-diabetics, minor tranquilizers, oral contraceptives, and so on (Sironi 1992, p. 162). The wide array of new drugs changed medical therapeutics, and doctors' prescriptions, substantially. By 1963, branded drugs made up 81 per cent of prescriptions compared to 47 per cent ten years earlier. Drug expenditures became the main driver of expenditures, resulting from the increased number of insured entitled to drugs (which went from 9.2 to 23.8 million in ten years), of prescriptions per capita (which increased from an average of 3.4 in 1950 to 10.4 per eligible in 1963), and in their unit cost, which more than doubled (Bellioni 1964). Prescriptions by doctors paid fee-for-service were one-third higher than those on a lump-sum basis (Pedone 1973). Pharmaceutical expenditures were also the main target of cost control policies. INAM unsuccessfully asked the government for substantial drug co-payments and autonomously introduced a Therapeutic Formulary. INAM obtained a price discount of no less than 17 per cent from the industry and the authorization to distribute drugs at its own facilities, a strategy for saving on pharmacy surcharges widely adopted also much later.

Hospital expenditures began their overwhelming increase in the mid-1960s and exploded in the 1970s. Hospital care had essentially remained on the fringes of the intense social and economic development of the 1950s up to the first half of the 1960s. The main determinant of the growth in hospital spending in this early period was expanded coverage, whose economic impact on health expenditures was offset by low admission rates and low day rates, thanks to low wages and poor provision of qualified staff and technologies. In 1963, hospital spending equalled pharmaceutical expenditure for the first time at around 30 per cent of total INAM expenditure. From then on, hospital expenditure increased at ever-accelerating rates. In 1973, hospital admissions accounted for 57 per cent of INAM expenditure, while drug costs shrank from 30 to 22 per cent, the effect of drugs' price discounts and the partial absorption of drugs' prescriptions by increased hospitalization (Fig. 6.1).

The trend in hospital expenditure shows a specific dynamics. Between 1965 and 1969, increased day rates explain 60 per cent of expenditure growth and frequency of hospital admissions explain 33 per cent, while the length of hospital stay remained stable at thirteen days. A negligible increase in the number of insured persons accounted for just 7 per cent of total expenditure growth (Iuele 1976). In the 1970s, hospital admissions increased by 52 per cent and day rates tripled, mostly because of increased staff expenditures. The huge size and swift dynamic made hospital expenditures the main source of financial instability of sickness funds and the cause of their recurrent financial crises, which prompted repeated state interventions. This requires a closer look at the process of modernization of Italian hospitals in the years of the economic miracle and at the time of their reform in the 1970s.

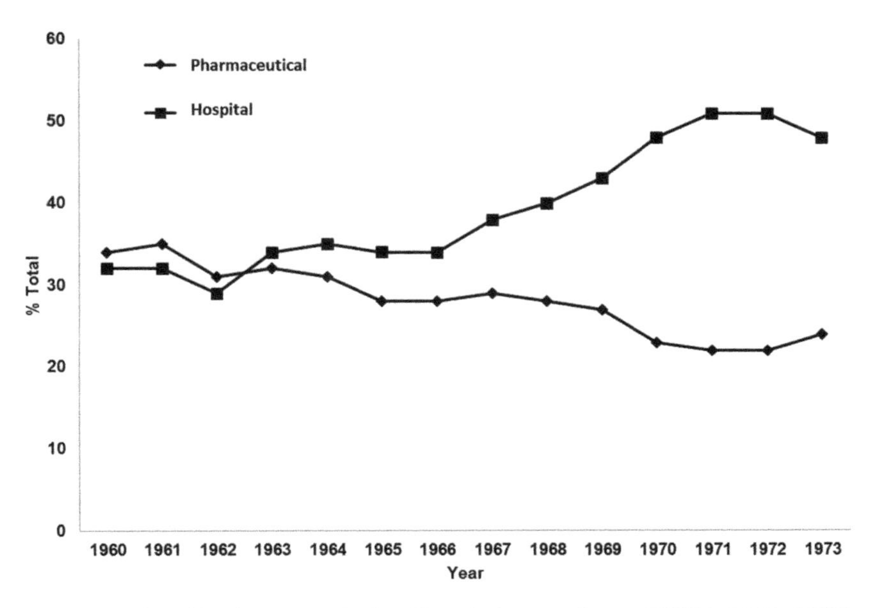

Fig. 6.1 Trend in pharmaceutical and hospital expenditures, INAM, 1960–1973. (Sources: Annali di Statistiche Sanitarie INAM, various years)

Hospitals: Insufficient, Inadequate, and Unprepared

At the end of the war, Italian hospitals were in a "very defective situation" (Cramarossa 1947). They clearly appeared inadequate to meet the current needs of the population, unprepared for the technological and organizational modernization imposed by postwar medical developments and to changes prompted by the expansion of insurance coverage and the new social demands. The first comprehensive survey of the Central Statistical Institute of 1954 identified 2315 hospitals, including sanatoria and asylums, with 362,053 beds (Istat 1955). The nationwide distribution at a rate of 7.5 per 1000 population was extremely skewed with 10 beds per 1000 in the north, 8.3 in the centre, and 4.3 in the south. Of the 1422 public general hospitals, 439 were simple infirmaries and 480 third category hospitals, poor in medical and nursing personnel and concentrated mainly in the southern regions, particularly in Basilicata and Calabria. Infirmaries averaged forty beds attended on average by two doctors and four nurses, only half of whom were licensed. Fifty-two class one and specialized hospitals, mostly concentrated in the large northern cities as well as Rome and Naples, averaged 1100 beds and were staffed by one doctor for every eight beds and a nurse for every five beds (Table 6.2).

One of the fundamental problems with hospitals in the first twenty years of the postwar period was the number of doctors and their payment. With the exception of the head of clinic (*primario*) appointed by the Boards of Directors, his *aiuti* (helpers) and *assistenti* (assistants) under the collective name of *secondari* (secondaries) had contracts lasting a maximum of four years for *aiuti* and two years for *assistenti*, renewable only once, and drew most of their income from *casse mutue*. *Casse mutue* paid a fee for each admission of their insured population; *primario* took half of the fees while the rest was divided among his secondaries. Moreover, a host of *assistenti volontari* (volunteer assistants) worked for free, in exchange for some training and reputation, and particularly, opportunities for private clients and, more often, *mutuati* to enrol in their list. Concerns about the number and qualification of nursing staff were widespread and led municipalities to organize active recruitment campaigns in the countryside and to enter into agreements with religious orders.

Hospital expenditures for the care and maintenance of their patients were covered by day rates paid, generally late and often after many disputes, by sickness funds, municipalities, and, to a limited extent, the patients themselves. By 1948, *casse mutue* had supplanted municipalities as the main hospital payers and provided 54 per cent of hospital revenues. The proceeds from self-paying patients made up just 2 per cent of total revenues (Grassini 1949). Day rates were set by the hospital board according to rules set by the Ministry of the Interior and included only direct

Table 6.2 Distribution of public hospitals according to Petragnani categories, 1954

Category	*Hospitals*			*Beds per*	
	Nr.	*%*	*Mean N. beds*	*Physicians*	*Nurses*
I	52	4.8	1170	8	5
II	100	9.3	374	12	7
III	480	44.8	123	19	9
Infermerie	439	41.1	40	14	9
Overall	1071	100.0	164	12	7

Source: Cherubini 1961

costs. As depreciation and maintenance expenditure could not be charged, hospitals had to self-finance their development through return on their own patrimony. However, the costs of new construction, expansion, and modernization of existing hospitals and the acquisition of the expensive equipment produced by medical developments in the early postwar years were far beyond the self-financing possibilities of individual hospitals. Particularly affected were once again southern hospitals, which further increased the imbalance in the hospital system. In the 1950s, hospitalization rates were low even among persons insured with INAM (Fig. 6.2).

To explain the low number of hospital admissions, some theorized a *ripulsa secolare* (centuries-old enmity) of the Italian population against hospitals, particularly in the south (Spinelli 1958), and campaigns were organized to advertise hospitals' "free and comfortable stay and good food" (Renda 1959). To others, it was increasingly obvious that hospitals were inadequate, could not keep up with the cost of rebuilding, acquiring modern technologies, and paying the expanded staff of doctors and licensed nurses they needed. Piero Grassini complained that the hospitals'

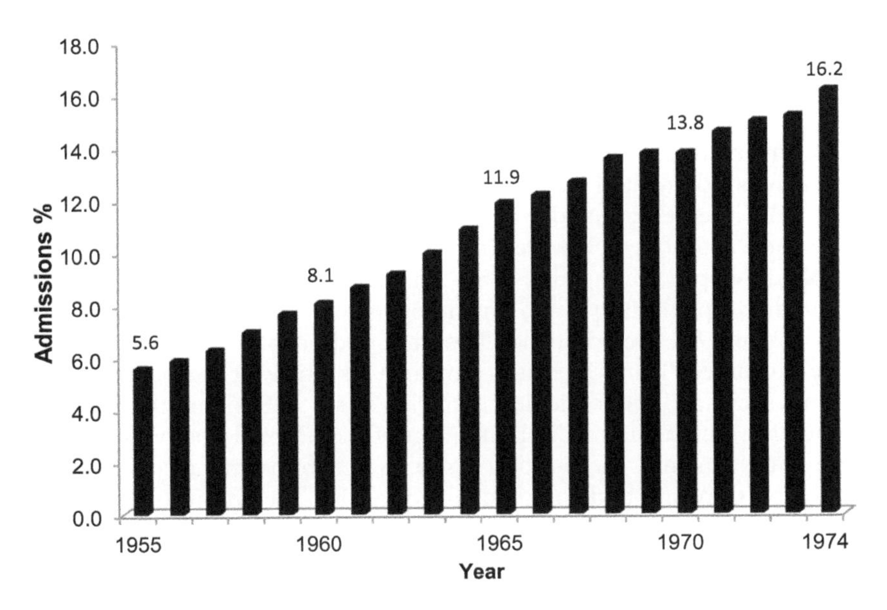

Fig. 6.2 Percentage of hospital admissions among persons insured with INAM, 1955–1974. (Source: INAM, Annuario Statistico, various years)

cherished autonomy had degenerated into "isolationism" and was pushing each hospital "to realize itself as a small world as complete as possible" preventing the establishment of "the hierarchical relationships among hospitals that a modern system required" (Grassini 1949). The National Congress of Social Medicine held in Bologna on November 1956 developed these same concepts (Ragazzi et al. 1956). The Congress proposed a "National plan for hospital care" based on the "chain system" (*sistema a catena*) where each hospital would be associated with "a specific range of action and its own territory, vertical and horizontal". The vertical dimension defined the relationships between hospitals at various levels of specialization, while the horizontal dimension ensured the integration between domiciliary and hospital care. This would ensure the free movement of patients according to their care needs, beyond insurance regulations, and make "all citizens equal in the face of hospitalization" (p. 10). This would also conform to "the developments of modern medicine", which required a re-conceptualization of the hospital based on its functional and technologic capacities "and not just on its physical capacity". The chain system shows many analogies with the concept of "hierarchical regionalism", which was informing both the hospital reconstruction programme in the United States with the Hill-Burton Act and the reorganization of the British hospital system on the model of the Emergency Medical Service (EMS) created during the war (Fox 1986).

A remarkable agreement emerged that hospitals were unable to meet the demands of medical modernization and their social function depended on public intervention and a new organization, which would necessarily limit their autonomy. However, the concepts that emerged and the proposals developed during the 1950s and early 1960s were never implemented. Hospital policy was a "mending policy" (*politica del rammendo*) (Cherubini 1961) based financially on the traditional, Giolittian policy of state-guaranteed loans. Hospitals remained essentially at the margin of the intense social and economic developments of the 1950s until the pressure of public opinion and medical unionism of the hospital doctors suddenly brought them into the political arena as a national political question. Public pressure and trade union demands pushed the hospital question to the centre of the national political agenda, where they met the "culture of planning" to carry out "structural reforms" of a series of governments of the centre-left (Ginsborg 1990, p. 210). The crisis of Italian hospitals was a crisis of relative scarcity in a period of increased consumption and economic expansion. The general improvement in living conditions brought about by the economic boom of the late 1950s had profoundly changed customs and lifestyles, increased individual consumption, and projected new needs for collective services. Massive migration from the south to the north of the country and from the countryside to the city had overcrowded and undermined even the large and (relatively) wealthy hospitals of the industrial cities of the north, also making their bed supply inadequate for the ever-increasing population. Moreover, expansion of health insurance coverage increased the demand for health services. The sudden disappearance of "the centuries-old enmity against the hospital" was channelling patients towards hospitals, now seen as the central location of modern medicine.

A series of reports on the state of the hospitals in major Italian cities between 1963 and 1964 by the weekly *Il Tempo* had the tone of muckraking, aiming to raise public awareness about the poor conditions of hospitals and other health services (Giannelli and Raponi 1965). Hospital conditions were defined a "civil shame" caused by the neglect of the state, whose disengagement was stigmatized by the headline claiming that "205,000 beds are missing and 7,000 million lire are needed". Complaints focused on the shortcomings of the hospitals in the south and the overcrowding of the large hospitals in the north, resulting from the emigration of the southern population to northern industrial centres. Hospital doctors were represented as the true bearers of medical science, and their poor condition was contrasted to the hasty medicine of *medici della mutua*, contributing to make hospitals the modern centre of scientific medicine.

The National Association of Hospital Aids and Assistants (*Associazione Nazionale Aiuti e Assistenti Ospedalieri*-ANAAO) was founded in 1959, the first union of hospitals' *secondari*. It quickly established itself as a national political force and played a fundamental role in the 1960s' reform of the legal status and economic treatment of hospital doctors, and of the general hospital reform of 1968 (Perraro 2009). This contributed to the fragmentation of the medical profession, as the unionism of *secondari* clashed with National Federation of Orders of Doctors (*Federazione Nazionale degli Ordini dei Medici*-FNOM), which claimed exclusive representation of all Italian doctors. Within the hospital sector, a further division ran by specialty and rank, between *primari* and *secondari*. The former was particularly between the doctors attending the wards and those in specialist services such as X-rays and labs but also medical specialties as cardiology, which were increasingly becoming autonomous from the large, self-sufficient divisions of "internal medicine" and "general surgery" that continued to characterize most Italian hospitals. The union's demands and the pressure of public opinion found a favourable reception in the new centre-left coalition, which made the "policy of planning" its distinctive political element for carrying out "structural reforms". Healthcare was the ideal object of application of the tools of the "culture of planning" to guide the action of the state towards the promotion of collective consumption and the overcoming of territorial imbalances, particularly manifest in the hospital setting. Health and the hospital question, in particular, became a central issue in the political agenda. In the elections of 1963, the Socialist Party adopted the slogan "a home, a school, a doctor for all". The socialist Mancini, Minister of Health in the first centre-left government, put the hospital as "the basic structure of the health system, the pivot of the entire health care system" (Mancini, cited in Giannelli and Raponi 1965). This opened the way for hospital reform as the first step to health reform.

As usual with health legislation, it took more than five more years, and several Commissions and Study groups, to pass the hospital reform bill, but the 1968 reform of Italian hospitals was, in many of its parts, largely innovative. According to Elio Guzzanti, the reform transformed the hospital into a public organization that "belongs to everyone and is open to everybody" (Guzzanti 1972). This focused on the universal right of access to hospital and the participation in its governance of representatives of the community of reference as well of its staff. Article 1 of the Act declared access to hospital care "a right for all persons present in the Italian territory", whether they were Italian citizens or from foreign countries. The cumbersome bureaucratic procedures to check patients' eligibility had to follow admission, which was a clinical decision of the doctor on call. Hospital governance included the Health Council made up of head physicians and representatives elected by the aids and assistants and was chaired by the medical director, who also participated in meetings of the Board. Unlike Petragnani law of 1938, which classified hospitals according to their average daily census, the Mariotti law adopted the concept of "hierarchical regionalism" in the organization of care and provided for three-level hierarchical order (local, provincial, and regional) hospitals, based on hospitals' catchment area and service-mix. Provincial hospitals were to be the backbone of the new hospital system, holding about 600–800 beds and serve a population of around 100,000 persons, the average size of Italian provinces. The scheme was clearly similar to that envisaged by Enoch Powell's 1962 Hospital Plan in England centering around the District General Hospital (Cutler 2011), but no information is available on some kind of "social learning" from abroad is available. Moreover, the large general medical and surgical wards of up to 100 beds still quite common were to be split into specialized, autonomous units including clinical sections of cardiology, neurology, geriatrics, orthopaedics, and thoracic services. This accelerated the adoption of new technological applications and the qualification of the medical and nursing staff, which increased in number and expertise. As Rudolf Klein remarks about Powell's Hospital Plan, this was "the child of a marriage between professional aspirations and the new faith in planning" (Klein 2006, p. 54). However, the law encountered the opposition of the more conservative professional groups, which lamented that the new autonomous services caused "the debasement of the value of the division, which represents the fundamental hospital unit". Conflict with organized medicine also concerned the legal status and economic treatment of personnel. To prevent *impiegatizzazione* (clericalization) of doctors and the "nationalization" of hospital medicine, the binding National Contract negotiated between national representatives of doctors and hospitals was downgraded to a national framework followed by local negotiations at the hospital level.

The issues concerning the institution of a National Hospital Fund had implications of great institutional and political significance. State financing was the logical consequence of the universal claim of free access to hospital care declared by article 1 of the law: a service open to everybody had to be necessarily paid by everybody. It also allowed the direct intervention of the state in financing the development of organizations that had proved to be inadequate to meet the costs of technological and organizational innovation of modern medicine. State funding was also the fundamental link between hospital reform and general health reform. However, the institution of a National Hospital Fund with the Ministry of Health implied the concentration of powers still spread across a number of other ministries, as we have seen with the Ministry of "*salvo*". This was of great political relevance in coalition governments. Moreover, a new centralized funding source conflicted with the virtual monopsony power of sickness funds over hospital finances, and infringed on the alleged autonomy of hospitals, and the de facto control that municipalities had on them. This led to a wide coalition of interests against the institution of an effective National Hospital Fund, which was dwarfed in size and aims. It amounted to just ten billion lire intended "for the granting by the Ministry of Health of contributions and subsidies to hospitals" for the renewal of equipment "in cases day rates… fail to cover the necessary expenses", as the law explicitly prescribed.

Mariotti's law on hospital reform has attracted several criticisms, which started at the time of its first draft and continue to the present day. The strongest criticisms blamed the law for the explosion in health expenditure observed between 1970 and 1974, contributing to the rising Italian public debt (see, e.g., Salvati 2000). Such criticism deserves to be reconsidered in the context of the historical evaluation of the modernization process of the Italian hospital system. During the 1950s, the financing of health services made little allowance for the correction of the well-known problems in their distribution and standard of provision, nor did it keep abreast of the postwar revolution in health services. State intervention was minimal, and *casse mutue* were only interested in purchasing cheap services for their insured. At the beginning of the 1960s, the state of the hospitals and the working conditions of their staff were the "civil shame" denounced by *Il Tempo*. Drugs were the main driver of health expenditure until the mid-1960s, a condition uncommon in most other OECD countries, where hospitals were the most expensive part of the health system. This was a sure sign of the backwardness of the Italian hospital system. The accelerated growth in hospital spending that began in the second half of the 1960s should be interpreted as the effect of the belated modernization of the Italian hospital system. The catch-up of the poor status of Italian hospitals with the expectations of the general public as well as of the medical staff prompted the interventions of the second half of the 1960s, which peaked with Mariotti's law. The political economy of the rise of hospital care costs can therefore be interpreted according to the classic "Weisbrod quadrilemma" of the interaction between technological change, increased quality of care, and demand for insurance coverage, resulting in cost containment (Weisbrod 1991). The expansion of insurance coverage in the 1950s improved the access of an increasingly large fraction of the population to increasingly effective and expensive services. The increase in health services usage, also encouraged by the economic boom, in turn stimulated the demand for new services and the further expansion of insurance coverage to meet their increasingly high costs.

A Legitimacy Crisis

Advances in medical knowledge and rising public expectations placed ever-increasing demands on health services and their standards. Sickness funds were scrutinized for their financial performance, but also for administrative capabilities and the quality of their services. In the early 1960s, while Italian gazettes were focusing on hospital miseries, a series of articles by the British weekly the *Economist* chastised the dubious practices and exorbitant power of the sickness funds. The *Economist* noted that "with great satisfaction, Italian authorities boast of the fact that sickness funds provide for the health of ninety per cent of the population. In fact, they do much more than that. An investigation by the *Ministero della Sanità* performed at the request of the Economist [...] reveals that the number of people insured by the eleven largest sickness funds amounted to 68,427,122 units, about sixteen million people in excess of the official national population". However, "*casse mutue* are powerful and immensely rich, and a position in their bosom is one of the most coveted prizes that political parties use to distribute. It will take more than a few scandals to destroy an administration system in the hands of public bodies, which are however alien to governmental control" (Jucker 1965, quoted in Giannelli and Raponi 1965, p. 107).

The legitimacy crisis also involved the quality of their services and called into question the very competence of the *medici della mutua*, who were ridiculed by a series of very successful movies by Alberto Sordi adapted from D'Agata's book *Il medico della mutua* (D'Agata 1964). Doctors opposed *casse mutue* since their institution with the Labor Charter. The arguments heard in the 1930s resonated also in the 1950s and 1960s in the speeches of the professional elite and were the cornerstones of the policy of the *Federazione Nazionale dell'Ordine dei Medici* (FNOM—National Federation of Orders of Doctors). Expanded mutual coverage guaranteed a flow of financial resources to the healthcare market resulting in its stabilization and increased occupational prospects. On the other hand, this also eroded private demand, and the growing number of doctors had to compete for a declining number of private patients and even for *mutuati* to enrol in their list. Extending insurance coverage to categories of workers, which were supposed to have the means to pay on their own, led to the same complaints that had met the coverage of sharecroppers in the mid-1935s.

Commenting on an article in the medical press significantly titled "the doctor on the cross", Giovanni Berlinguer noted that "each new category, each new social group gaining health coverage under a new law is but one more nail, one more torment for the free profession of doctors" (Berlinguer 1957). Doctors lamented that the intrusion of a third party between the doctor and his patients hindered the establishment of an "intimate dialogue". In the long term, it would also lead to "the degeneration of medical practice and medicine, undermining not only freedom and dignity of the doctor but above all the quality of medicine" (Ilardi 1964). This induced, particularly among the professional elite, the regret of an imagined past: "if there were a country in the world so well organized in healthcare that nobody was in want of medical care, this was Italy. *Medici condotti* took care of the poor, and hospitals were obliged to take them in. And we had municipal assistance. Why did we need to bring in this *mutue* in addition to municipal assistance, which it's no use except in keeping despicable parasites among us?" (cited in Secco 1996, p. 199).

The Reformist Mirage

The Plan for Economic Development for the years 1965–1969 prepared by two consecutive socialist Treasury ministries, Antonio Giolitti and Gaetano Pieraccini, laid out the outline of a state-financed national health service managed by local authorities which made a significant break with the minimalist, "mending policies" of the 1950s. They both failed, although in different ways. Giolitti's plan was not discussed, while Pieraccini's plan was formally enacted but was not implemented. They however left a significant cultural impact on health policymaking. Throughout the 1960s and the first half of the 1970s, two competing health policy proposals opposed each other, proposing either structural reform or simple reorganization.

The Plan for Economic Development set out three goals for the health sector. The first was "to ensure comprehensive health protection for the whole population" and the second "to increase the preventive nature of health interventions", a traditional shortcoming of casse mutue. The third was all-encompassing and aimed "to do away with the insurance principle in healthcare and set up a national health service which is financed by the citizens in proportion to their income, and managed by the State through the Ministry of Health, the Regions, Provinces, and Communes". The Plan also provided for the establishment of Local Health Units, over 200,000 additional hospital beds, of which half would be in the south, a hospital network hierarchically divided into basic hospital units, main hospitals, and regional hospitals. This required 50,000 additional doctors, of which 7000 in the first five-year period of the Plan. Expenditures were estimated at 5300 billion lire, to be financed in the main through savings from pharmaceutical expenditure.

The opposition came in the main from *casse mutue* and the National Federation of Doctors' Orders (FNOM), which found in a common enemy the reason for strategic appeasement. The main inspiration for a simple reorganization of social security came instead from the National

Council for the Economy and Labor (CNEL). Its "Observations and proposals on the reform of social security" (CNEL 1963) included forty "conclusions" in the form of short sentences inspired by the conclusions of Commissione D'Aragona, of which Coppini, speaker of the Report and president of INAM, had been a member. The CNEL proposals for the health field included state financing of public health services and hospital care, which should be "opened to all citizens". Hospitals maintained their autonomy, but provision was to be brought under the control of *casse mutue* participating in their boards. Sickness insurance was to be "general", including insurance against tuberculosis, and *casse mutue* would further increase their network of owned and managed outpatient specialist clinics. As far as the old issue of the fragmentation of *casse mutue,* their unification was deemed "excessive", and some "functional grouping" was envisaged into two and possibly three main *Enti nazionali* (CNEL 1963).

FNOM countered governmental plans to establish a national health service offering *casse mutue* an alliance against the looming common threat of "statization" of medical care. The strategic alliance envisaged doctors accepting the system of *casse mutue* on condition of their respect of political and economic dimensions of medical dominance. FNOM asked *casse mutue* to cover only the "economically weak", allow doctors free access to the market of the insured through open panels, and adopt fee-for-service and indirect form of payment to protect "liberal" medical practice. The instrument of the compromise was the "Scheme of a bill for the reform of social insurance of illness prepared by the National Federation of Doctors' Orders" drafted by a joint committee chaired by the president of FNOM Raffaele Chiarolanza (Chiarolanza 1963). The unrelenting support of the existing system and unwavering opposition of governmental reforms contributed to the fragmentation of the medical profession and the mushrooming of sectorial organizations. As we have seen, the organization of hospital doctors actively supported hospital reform. The Italian Federation of *medici mutualisti* (contracted with *casse mutue*) (FIMM) held its national congress in Bologna under the slogan "a general reform of healthcare is mature and urgent" (Vicarelli 2006). The FNOM countered the fracturing of the medical class by placing itself as the clearinghouse of special interests in the name of the superior, and therefore necessarily common, ethical, and deontological principles of the profession. Special interests and general principles however clashed when FNOM claimed the exclusive representation of the profession in labour negotiations. This proved short sighted and contributed to the further fragmentation of the medical front.

The fragile equilibrium that the weakened forces of *casse mutue* and organized medicine had struck with a divided government sustained a stalemate that could have been long lasting if it had not been broken by a combination of endogenous and exogenous factors. The dynamics triggered by the modernization process of the hospital increased hospital expenditure swell above the financial capacity of the sickness funds. Between 1967 and 1974, three legislative decrees authorized ten-year loans for a total of over 3500 billion liras to finance the main *casse mutue* whose enormous debt exposure to hospitals was threatening their survival. The financial crisis of 1967 led to additional state funding for INAM, ENPAS (covering state employees), and *Coldiretti* (covering farmer proprietors) totalling 476 billion lire. The deficit of 117 million lire of *Coldiretti* compared to INAM's 280 million was particularly troubling as the number of its policy holders were less than one-fifth the size of those with INAM. The recurring cost overruns of sickness funds suggested structural, not cyclical problems, and prompted pervasive concerns about the viability of relying on employment and payroll contributions as the basis for sickness insurance. In the years of the economic miracle, an expanding labour market and increasing wages had offered a solid base for developing a system of health insurance based on employment. However, by the end of the 1960s, changes in the labour market and in society made employment a less suitable basis for social insurance, whose inherent inequalities appeared less and less socially acceptable. But it was, above all, the political initiative of social movements and the institutional activism of the regions which decisively shifted the balance towards a wholesale reform of the health system.

References

Bellioni, M. (1964) L'assistenza farmaceutica dell'INAM dal 1950 al 1963. I Problemi della Sicurezza Sociale, 19 (6), pp. 905–937
Berlinguer, G. (1957) La medicina si trasforma: Nuove responsabilità del medico. Rassegna Sindacale, 6–7, pp. 171–173
Berlinguer, G., Delogu, S. (1959) La medicina è malata. Bari, De Donato
Berlinguer, G. (1968) Economia della salute [1966]. In: Berlinguer G., Sicurezza e insicurezza sociale. Roma, Leonardo, pp. 239–271

Brenna, A. (1974) L'economia del sistema sanitario italiano. Le Scienze, 73, pp. 110–122

Bruzzese, U. (1967) (a cura di) La situazione sanitaria assistenziale e previdenziale in Italia. Roma, Istituto Editoriale Scientifico

Cherubini, A. (1961) Aspetti organizzativi, economici, sanitari dell'assistenza ospedaliera. Roma, Istituto Italiano di Medicina Sociale

Cherubini, A. (1977) Storia della previdenza sociale. Roma, Editori Riuniti

Chiarolanza, R. (1963) I secolari principi della libertà professionale. Federazione Medica, 16 (4), p. 5

CNEL—Consiglio Nazionale dell'Economia e del Lavoro (1963) Osservazioni e proposte sulla riforma della previdenza sociale. Assemblea 3 ottobre 1963, Roma

Colombo, V. (1977) La giungla sanitaria. Idee per una riforma dei servizi sanitari. Torino, Vallecchi

Coppini, M.A. (1959) Che cosa ci separa dalla sicurezza sociale. I Problemi della Sicurezza Sociale, 14, pp. 173–187

Crainz, G. (2003) Il paese mancato: Dal miracolo economico agli anni ottanta. Roma, Donzelli

Cramarossa S. (1947) La ricostruzione sanitaria: Igiene e Sanità Pubblica, 3, pp. 266–276

Cutler, T. (2011) Economic liberal or arch-planner ? Enoch Powell and the political economy of the Hospital Plan. Contemporary British History, 25, pp. 469–489

D'Agata, G. (1964) Il medico della mutua. Milano, Feltrinelli

Delogu, S. (1967) Sanità pubblica, sicurezza sociale e programmazione economica. Torino, Einaudi

Fox, D.M. (1986) Health policies, health politics: The British and American experience, 1911–1965. Princeton, Princeton University Press

Giannelli, G., Raponi, V. (1965) Libro Bianco sulla riforma ospedaliera. Roma, Supplemento del notiziario dell'amministrazione sanitaria.

Giorgi, C., Pavan, I. (2021) Storia dello Stato sociale in Italia. Bologna, Il Mulino

Ginsborg, P. (1990) A history of contemporary Italy: Society and Politics, 1943–1988. London, Penguin Books

Grassini, P. (1949) Il problema degli ospedali in Italia. Cronache Sociali, 6, pp. 5–9

Guzzanti, E. (1972) Problemi di funzionalità interna dell'ospedale nelle prospettive della Legge di riforma: Atti prime giornate di studio su L'Ospedale in Italia: Strutture e Funzionalità. Roma, Istituto Italiano di Medicina Sociale, pp. 19–34

Klein, R. (2006) The new politics of the NHS: From creation to reinvention. Oxford, Radcliffe Publishing

Ilardi, V. (1964) Libertà, dignità e qualità della professione medica nella mutualità. Federazione Medica, 2, pp. 38–41

INAM News (1959) la "piccola riforma" dell'INAM. I Problemi della Sicurezza Sociale, 14, pp. 341–357

Istat—Istituto Centrale di Statistica. (1955). Annuario Statistico Italiano. Roma, Istituto Poligrafico dello Stato, pp. 37–56

Iuele, R. (1976) L'assistenza ospedaliera INAM nel periodo 1950–1974. I Problemi della Sicurezza Sociale, 31, pp. 495–525

Lanaro, S. (1997) Storia dell'Italia Repubblicana. Venezia, Marsilio

Melis, G. (1996) Storia dell'amministrazione italiana, 1861–1993. Bologna, Il Mulino

Papa, G. (1958) Sull'andamento della mortalità e della morbilità tubercolare in Italia nel campo della protezione assicurativa. Previdenza Sociale, 14, pp. 51–84

Pedone, A.P. (1973) Ricerca sull'andamento territoriale delle prescrizioni farmaceutiche. I Problemi della Sicurezza Sociale, 28, pp. 289–313

Perraro, F. (2009) La nascita dell'ANAAO. In: I 50 anni dell'ANAAO-Assomed, 1959–2009, Roma (pp. 25–60)

Petrilli, G. (1953) La sicurezza sociale. Bologna, Cappelli

Ragazzi, C.A., Petrilli, F.L., Moretti, I., et al. (1956) Mozione conclusiva. In: Società Italiana di Medicina Sociale. Atti del VI Congresso Nazionale di Medicina Sociale, Bologna, 1956. Milano, Minerva Medica, 1958 (VII–VIII)

Renda, C. (1959) Per una campagna propagandistica per lo sviluppo dell'assistenza ospedaliera in Sicilia. Difesa Sociale, 38, pp. 84–95

Salvati, M. (2000) Occasioni mancate. Economia e politica in Italia dagli anni '60 ad oggi. Bari, Laterza

Savoini, V. (1955) Intervento. In: Atti Convegno Nazionale ACLI. Roma, pp. 38–43

Scarpa, S., Chiti, L. (1975) Di farmaci si muore. Roma, Editori Riuniti

Secco, A. (1996) La corporazione medica in età repubblicana. In: Storia d'Italia, Annali 10. Torino, Einaudi, pp. 193–221

Sironi, V.A. (1992) Le officine della salute. Storia del farmaco e della sua industria in Italia. Bari, Laterza

Somogyi, S., Pennino, C., Vizzini, S. (1974) L'assistenza sanitaria comunale 1952–1970. Roma, Istituto Italiano di Medicina Sociale

Spinelli, O. (1958) Sull'assistenza ospedaliera in Italia specie nel mezzogiorno e nelle Isole. I Problemi della Sicurezza Sociale, 3, pp. 357–373

Tronelli, M. (1973) Evoluzione dell'attività normativa dell'INAM. I Problemi della Sicurezza Sociale, 28 (3), pp. 607–632

Vicarelli, G. (2006) Medicus omnium. La costruzione professionale del medico di medicina generale, 1945–2005. In: Cipolla C., Corposanto C., Tousijn W., Eds. I Medici di medicina generale in Italia. Torino, F. Angeli, pp. 50–99

Weisbrod, B.A. (1991) The health care quadrilemma: an essay on technological change, insurance, quality of care and cost containment. Journal of Economic Literature, 29, pp. 523–552

CHAPTER 7

The Creation of the Servizio Sanitario Nazionale

The "long" Italian 1970s spanning from the late 1960s to the early 1980s were one of the most intense and contradictory periods in the life of the country, masterfully represented by Alberto Arbasino as "the decade of illusions, of the fall of illusions, of the dramatically decreasing distance between the rise and fall of illusions" (Arbasino 1980, p. 9). In the wake of the economic miracle, workers' and students' *autunno caldo* (hot autumn) was a period of social unrest and political turmoil, "the high season of collective action" (Ginsborg 1990, p. 298). The early 1970s saw the expansion of social and civic rights in several areas, including pensions, public housing, labour, school, family, divorce, and abortion. A major step in the transformation of the structure of the state was the introduction of regional governments in the spring of 1970, twenty-two years after their provision in the Constitution. Two oil crises abruptly halted economic development and interrupted the "golden age" of the welfare systems. Stagflation, that is, low growth, high inflation, and high unemployment threatened a "fiscal crisis of the state" (O'Connor 1973) and triggered a "financial and fiscal imperative" which focused on curbing welfare states which were perceived as having grown "beyond their limits" both in expenditure and in bureaucracies (Habermas 1976).

The economic crisis hit Italy very hard and peaked in 1974–1975, when inflation reached 20 per cent and many believed that "the authorities had almost lost control of the situation" (Spaventa 1988, p. 10). The "Season of the movements" of women, workers, and students degenerated in the

F. Taroni, *Health and Healthcare Policy in Italy since 1861*,
https://doi.org/10.1007/978-3-030-88731-5_7

anni di piombo (years of lead) of political terrorism. Society accentuated its fragmentation and individualism in what came to be known as *il riflusso* (the ebb) or the *età dei cespugli* (age of the bushes) after the image of Censis, which marked the rise of "a logic of individualism in which everything exists except collective morality, civil conscience, sense of the institutions" (Censis 1979). In the wake of the kidnapping and killing of Aldo Moro, the Communist Party voted a government of *solidarietà nazionale* (national solidarity) for the first time since 1947. This was the difficult economic and political context within which a heterogeneous coalition in 1978 passed with a landslide majority Law nr. 833 instituting the *Servizio Sanitario Nazionale* (SSN—National Health Service). A law that significantly expanded social rights and possibly increased public spending while the other countries were adopting restrictive new policies of retrenchment (Pierson 1994, 1996).

The Collapse of the Sickness Funds

Casse mutue had successfully resisted timid attempts at reform in the 1960s, despite their recurring cost overruns, estimated at 467 million lire in 1967 and 400 million lire in 1970. The situation changed suddenly with the conversion into law of a decree to remedy their third financial crisis in a decade with deficits that were initially estimated at around 2.7 billion lire but would later prove to be over 4 billion lire. The "August battle" waged in the Parliament in August 1974 against the law opened a window of opportunity which suddenly and somewhat unexpectedly turned what had been for decades "a reform difficult to pass" into "a reform impossible to avoid" (Berlinguer 1979). The title of Law nr. 386 of August 17, 1974, carried three political and institutional breakthroughs: "Rules to pay off *casse mutue* debts towards hospitals, finance hospitals' expenses and start health reform". This announced the very political decisions to let sickness funds go bankrupt, institute a national hospital fund (Minister Mariotti's impossible dream in 1968), and transfer hospital planning to the regions. These provisions were the explicit repeal of the social insurance system and significant material steps towards a general reform of health organization. In the meantime, the Minister of Health Vittorino Colombo presented to the Parliament the government bill for the establishment of the SSN. It would take four more years before the final approval, but change had been set in motion. Its agents were primarily the regions and social movements.

The Regions, Laboratories of Democracy

While the financial crisis had acted as a catalyst, the "unfreezing" of the Constitution with the institution of the regions in 1970 and the transfer of state functions in the field of "health and hospital assistance" were critical to overcoming the deadlock that had been blocking change in the organization of healthcare for thirty years. This was also a significant opportunity for the regions. Administrative functions had been transferred only in bits and pieces and with strict financial constraints. Taking responsibility over hospital planning and the regional hospital budget and undertaking the management of health services from *casse mutue* implied a significant empowerment of the regions in the politically relevant field of health and healthcare. Moreover, the reform bill promised them a powerful position in the national political arena.

The regions unanimously urged the national government for fast approval of the health reform. Some of them used their legislative power to pass laws implementing innovative services, frequently born out of the experiences of local authorities and social movements. An intense process of social learning and technology transfer developed, where local "experiments" initiated in one region were adopted by others controlled by different political majorities and, at times, transfused into national legislation. This is the case, for example, of *consultori familiari* (women's clinics, of which more below) and the social experimentation of *consorzi socio-sanitari* (area social and health consortia) between municipalities for the associated management of environmental, occupational, and public health services in Tuscany, Emilia-Romagna, Veneto, Lombardy, and Puglia (Taroni 2015). In their early years, several regions were open and responsive to the demand and practices of social movements and showed innovative capacity in terms of organizational models and legislative production, even in the face of the distrust of the movements and the resistance of a recalcitrant central bureaucracy. This brought the Italian regions closer to Louis Brandeis' concept of "workshops of democracy" where American states "experimented with new social and economic policies without jeopardizing the rest of the country" and substantiates Putnam's conclusion that the regions were "the most significant innovation of Italian politics in more than thirty years of the Republic" (Putnam et al. 1983).

The shrewd political management of the umpteenth financial crisis of sickness funds and the strong presence of the new agents in the national political arena opened a window of opportunity for the first step towards a reform that had been repeatedly announced and continually postponed for over a decade. In addition, social movements gave a decisive boost to the process and contributed crucial ideas.

The Season of the Movements

The sweeping movements of social protest and political mobilization of the 1960s and 1970s frequently had health at their core. Health movements combined attention to the international circulation of ideas and experiences and the autonomous organization of self-managed, alternative practices with pressure from below to set the political agenda and demand institutional change. Trade unions frequently acted as a medium for the institutionalization of the issues raised by the movements, and the institutions often responded.

An example of the unusual openness of institutions to practices adapted and reworked from foreign experiences is the women's movement which developed in Italy "in the fabric of politics" (Ergas 1986). It combined its focus on the women's liberation and gender relations with social transformation, and engaged in direct confrontation with the medical establishment and the political system. With the campaign against the abortion act approved by the Parliament and for its "liberalization", that is, that the decision be solely up to the woman and that the medical service be safe and free, the feminist movement came to represent one of the main points of reference for women, regardless of political affiliation (Ergas 1982). The confrontation strengthened the movement, and women's health centres (*consultori*) were set up in several cities. *Consultori* adapted international ideas and experiences (a common reference was the volume *Our body, Ourselves*, originally a self-help manual from the Boston Women's Collective), which were transformed into self-managed practices, allegedly alternative to traditional medicine (Boston Women's Health Book Collective 1971). Successful experiences led to legislation in several regions, and then to a national law, which attributed responsibility for women's services to *consultori materno-infantili*, which replaced the old ONMI services for women, mothers, and children.

The mobilization of workers also produced ideological and institutional change in occupational health practices and their institutions, which also evolved into a form of "labor environmentalism" (Barca 2012). Work safety evolved around what came to be known as "the workers' model" of inquiry into health hazards and work organization (Berlinguer and Biocca

1987). The model was centred on the concept of the workers' homogeneous group, that is, the evaluation of workplace hazards by groups of workers performing similar tasks and presumably exposed to similar work-related health problems (Reich and Goldman 1984). Risk maps (*mappe di rischio*) provided a profile of the distribution and severity of occupational hazards in a factory or a community and were the main tool to select and organize priorities of intervention. The manual published by the Federation of Metal Workers was the basic text both for workers' education and for industrial negotiations on occupational health and safety problems (Oddone and Marri 1977). Basic procedures and tools were also transposed into an article of Law 833, which laid out the organization of the new *servizi di medicina del lavoro* (occupational health services).

The 1960s and the 1970s saw a radical shift in the interpretation and treatment of mental illness which transformed the life of people with the illness and turned asylums into open "therapeutic communities" (Foot 2015). Franco Basaglia and the movement of *psichiatria democratica* (democratic psychiatry) he inspired were the main agents of the de-institutionalization of the care of persons with mental illnesses, along with a critique of the medical establishment and of the value of medicine more generally. These were the main intellectual strands of the collection of writings *L'istituzione negata* (The institution denied), which Basaglia edited in 1968 and became a staple in the cultural debate (Basaglia 1968). The ensuing law 180 or "*legge Basaglia*" was finally approved ten years later, in May 1978, and was subsequently subsumed under the national reform act a few months later, making mental health an integral part of the SSN (Babini 2009, pp. 241–292). The law allowed for the gradual closure of asylums and the development of a network of community services, although designing and developing alternative institutions proved more difficult than expected (Fioritti 2018). The debate reached an international dimension and became a term of reference for agencies such as the WHO but was also criticized, particularly in England (Jones and Poletti 1985, 1986).

Social movements in the field of health shared some common traits, essentially consisting in a strong social orientation and a radical critique of authority and existing institutions towards de-institutionalization, mutual and self-help, and alternative practices. In medicine, common references were the critical analyses of the role of medicine in modern society and its perceived failings, including Ivan Illich's radical criticisms (Illich 1974), McKeown's thesis (McKeown 1976) and political analyses of the commodification of medicine (Ehrenreich and Ehrenreich 1971). On a different but concurrent level, critical appraisal of medical practice due to the widespread diffusion of practices of unproven efficacy and to the failure to adopt interventions of proven efficacy was popularized by Cochrane's book, published in England in 1972 and translated in 1978 (Cochrane 1972). The book was part of the successful collection *Medicina e Potere* (Medicine and Power) edited by Giulio Maccacaro, a leader and founder of *Medicina democratica* (Democratic medicine) (A collection of his papers is in Maccacaro 1979). A rich and original production attentive to international cultural and scientific developments also derived from the peculiar intertwining between diseases of the misery and disasters of progress, captured by the title "Urban Malaria" of the collection of essays by Giovanni Berlinguer published in 1976. The anachronistic cholera epidemic, which hit Naples and Bari in 1973, and the very modern TCDD toxic cloud of Seveso, near Milan, due to the explosion of the ICMESA reactor (Centemeri 2006) highlighted both the different development between the north and the south and the uniformly poor conditions of public health in all parts of the country.

Passing the Law

Four years after the August battle that had officially started the final leg of the reform, after two governmental bills by Ministers Vittorino Colombo and Luigi Dal Falco, four parties' bills, and the unavoidable technical commission, on December 21, 1978, the Parliament passed Law nr. 833 "*Istituzione del Servizio sanitario nazionale*" with 381 votes in favour, 77 against, and 7 abstentions. The Minister of Health was Tina Anselmi, the first woman minister in Italian history, and the President of the Council of Ministers was Giulio Andreotti, in his fourth term as the prime minister. In one of the several occasions when approval had seemed very close, Minister Labor and Social Welfare Luigi Bertoldi had predicted that the passing of the law would be "a moment of shouts and fury". Parliamentary debate instead dragged wearily "between a football game, a regional election and the election of the President of the Republic" remarked Deputy Luciana Castellina, while Deputy Giorgio Bogi found "illogical passing a reform in three days after a fifteen year delay". Except l'Unità and l'Avanti, the official organs of PCI and PSI, the daily press put the news in the inside pages. Newspapers' complaints about the poor performance of health services were taken as sure proof of the shortcomings of the reform. The speaker of the bill Danilo Morini couldn't but observe that no fault could bear upon a law yet to pass, but the four years elapsed since the alleged start of the reform were clearly taking their toll.

The institution of the SSN marked a radical break with the past, making healthcare available to all citizens on the basis of need rather than occupation and funded by centralized, progressive taxation instead of social insurance contributions. Extending insurance to all was an alternative that a wide and heterogeneous combination of egalitarian progressives and efficiency-minded technocrats considered less desirable. All but a few shared the view that the insurance system of *casse mutue* was a sickness, not a health, system, and a very unequal one. *Casse mutue* insured individuals against sickness, but did not ensure collective health in the community, the goal of WHO's Alma Ata declaration and of the Lalonde Report in Canada (Lalonde 1974). Moreover, sickness funds had become increasingly dependent on the public budget anyway, as contributions covered barely two-thirds of their expenditures, and the state had frequently intervened to cover their ever-increasing cost overruns (Brenna 1974). Financial problems and the changing structure of the labour market towards more flexibility convinced "rationalizers" that social insurance was not a choice. The SSN would not only solve the problem of selective coverage by introducing universalism. A centralized system of financing would ensure central government to maintain better control of health spending. They also hoped that global budgeting at the national level would be the lever to constrain providers to efficiency and cost containment (Rosaia 1976). The broad principles of universal and equitable coverage, public funding, comprehensive services, and local control were agreed upon by all political parties, except the Liberal Party (*Partito Liberale Italiano*) and the post-fascist *Movimento Sociale Italiano* (MSI), which voted against the law. Although the heterogeneous coalition supporting the law had differences over important issues, the reform included three fundamental breakthroughs in the history of Italian health organization.

The first period of the first article of Law 833 designated the SSN as the instrument to implement article 32 of the Constitution: "The Republic protects health as a fundamental right of the individual and in the interest of the community, *through the National Health Service*" (emphasis added). The literal transposition of article 32 in the text of the reform law marked its ultimate victory against article 38 that had legitimatized the mutual regime in the first thirty years of the Republic. This pulled health out of the particularistic and categorical fabric of the Italian social insurance system. Furthermore, the law clearly defined the Republic in its national, regional, and local tiers as the actor in implementing and running the SSN. This eliminated the central government's passive role of interest broker devoted to the resolution of periodic conflicts between sickness funds, doctors, and hospitals to envisage a direct responsibility of the state at all levels in financing, planning, and managing the new system. Finally, the law clearly established, for the first time, a variety of equity goals. The geographical distribution of hospitals and other medical services was grotesquely skewed towards the healthier and wealthier parts of the country because of centuries-old unequal development, which was superimposed over social differences in access to medical services brought about by the mutual insurance system. The law promised access to the same standard of service irrespective of income, work, and domicile (art. 3) and equitable standards of regional financing, with the additional goal of "progressively eliminating the structural and performance differences between the Regions" (art. 51). This would result in "uniform" health status (art. 4) and help "overcoming territorial imbalances in the social and health conditions of the country" (art. 2). However, the founding legislation envisaged no programme for re-distributing resources according to health needs.

At the organizational and managerial level, the law aimed at transforming the fragmented array born out of the sickness funds into an integrated and coordinated system. As with the British NHS, the SSN used a geographically defined community as a reference, ensuring the material basis for environmental prevention and processes of democratic participation, while vertical integration remedied the fragmentation of the mutual system. Local health units (*Unità Sanitarie Locali*—USL) were to take responsibility of promoting health and providing healthcare to populations living in defined geographical and administrative areas with between 50,000 and 200,000 residents. Moreover, the law set up a massive project of complete vertical integration on a territorial basis under the sole public ownership of the SSN of a wide range of collective programmes and personal services. These included general practice, outpatient specialist care, hygiene and prevention, occupational health and veterinary services. Hospitals lost their autonomy and became operational structures of the USL. About 15 per cent of inpatient care and about one-third of specialist ambulatory care remained private but had a contractual relationship with the SSN.

The functioning of the SSN envisaged a system of indicative planning at the national and regional levels and local managerial control by USL. The law established what came to be called a "cascade system" in which central government annually determined the national health budget within the three-year perspective established by the National health Plan (*Piano Sanitario Nazionale*—PSN). Proposals from the Treasury to set a ceiling at about 6 per cent of the Gross National Product "for the first year and in decreasing percentages in subsequent years" were unanimously rejected. The national budget was to be allocated to the regions based on their population, but also considering "the need to overcome the conditions of socio-sanitary backwardness of the country, particularly in the southern regions" (art. 53). The spending power of the central government's allocation was notionally entrusted to regions but the USLs did the actual spending.

Bugs and Holes of the Reform

According to Larry Brown, the Italian health reform was "an astonishing, indeed ostentatious display of political idealism" (Brown 1984). Law 833 was also a political and institutional compromise between central government, regions and municipalities, the Christian Democrats and the Communists, and many other interest groups, resulting in a sort of "a something for everyone reform package" (Brown 1984). This riddled the law with hype, hope, ambiguities, and contradictions, which led to some fundamental structural weaknesses of the SSN. The design of the SSN outlined by the law could plausibly be interpreted in two ways. The "cascade" model of central planning and financing designed a unitary organization hierarchically ordered on the three levels of state, regions, and municipalities. The principles of decentralization and participation of which the law was imbued allowed instead for a system of multilevel governance of autonomous institutions, whose harmonization was based upon the constitutional principles of solidarity, subsidiarity, and fair and loyal cooperation. In practice, the highly centralized tax reform of the early 1970s and the low tax base of the regions led to a financial system where spending at the regional and local level was decoupled from taxation. Lack of mechanisms to control cost and that of clear accountability for spending were the major weaknesses of the law, the not-so-obscure disease of which the reform was claimed to be "sick" (Cavazzuti and Giannini 1982). Responsibility for funding, which lay with the central government, was disconnected from responsibility for spending decisions, which rested with the USL, administered by the municipal governments.

The USLs were not just insulated from the financial implications of their spending actions. Their control by municipalities gave political parties opportunities for extracting rents in the form of patronage and, on occasion, for obtaining party funding from kickbacks (Golden 2003). The conditions were therefore ripe for the cost of the new system to grow rapidly. As imposing a ceiling on the national health budget had been rejected, users' co-payment was the tool available for balancing the budget. Furthermore, given the lack of designated intergovernmental institutions for coordinating health policies between regions and the central government and pending the approval of the first National Plan (which appeared only in 1994), fiscal policy was the main field of interaction between national and regional governments. National fiscal policies aimed at containing USL expenditures at the front end resulted in a "budget policy" (*politica di bilancio*), which remained a structural characteristic of the SSN for many years (Buglione and France 1983). The Treasury adopted a strategy of systematically underestimating projected expenditures and, hence, funding needs by overestimating the savings to be obtained from unrealistic cost containment plans. This created a situation of permanent fiscal crisis in the healthcare sector at the regional and local level that contributed to distraction from the obvious problems in implementing the reform. Budget policies and financial stringency had also a significant impact on the equitable goals of the SSN. Policies designed to remedy the inherited inequalities in the geographical distribution of services would necessarily have to transfer resources from the best-endowed areas to the more deprived regions of the country. Political expediency suggested instead an allocation based on inherited budgets, and this left undisturbed historical inequalities in the distribution of resources.

The issues about the legal nature of the USL and their optimal size were closely intertwined in the debate. The search for a balance between economies of scale and scope, political and professional representation, popular and users' participation proved unsuccessful. The compromise defined USL a "service complex" (*complesso di servizi*), legally *un oggetto misterioso* (a mysterious entity), as the parliamentary debate ironically defined it. The designated size of the USLs ranged between 50,000 and 200,000 inhabitants and was therefore essentially left to the regions to define. The decision to deprive hospitals of their autonomy was instead unambiguous, although a last-minute compromise. All these botched "solutions" were extremely fragile and remained recurring topics of a debate which peaked in the early 1990s with the corporatization of USLs and the *scorporo* (splitting) of major hospitals as independent corporations and are still alive today.

Along with choices over the governance of the USL there was the problem of the relationships with *Federazione Nazionale dell'Ordine dei Medici* (FNOM) and organized medicine more generally. The Parliament expressed a firm position for a political government of the USL and opposed to professional management by medical doctors at all levels of the new system. Resisting FNOM's pressures, technical bodies such as the regional and the local health council were erased from the government bill, and trade unions and professional representatives were excluded from the National Health Council. The governance of the USL was entrusted to a Management Committee, a political body made up of nominees from the political parties that held a majority in the councils of the constituent municipalities of the USL. The technical management of the USL was responsibility instead of a collegial body, the Office of direction, appointed by the Management Committee. The Office maintained a two-headed coordination, both medical and administrative, as FNOM's pressing request for a stand-alone medical director was ignored. This was just one of the manifold expressions of the "anti-technocratic and anti-elitist bias", which was widely diffused among "the public opinion and the political class itself" (Freddi 1990, p. 40). The strategy of FNOM, the most prominent institution of organized medicine, proved extremely flexible as demonstrated by the front pages of two consecutive issues of the weekly "Il Medico d'Italia", its official organ (Taroni 2011, pp. 209–210). The nine-column headline of the front page of the last issue of the year announced: "Reform without doctors". Appeals were made to Prime Minister Andreotti and President Sandro Pertini to withdraw their backing for the law. The press conference of the President of the FNOM, Eolo Parodi, voiced the strong opposition of the medical profession against the law. Among several issues, FNOM s lamented "the exasperated bureaucratic approach, the cumbersome management framework, the exclusion of the representation of doctors from all SSN bodies, the bureaucratization of doctors in general medicine, the lack of a medical role and the absence of protection for his professionalism". The next issue of the Medico d'Italia opened instead with a full page heading claiming "Health care reform: an event of great economic and social importance". A warning however followed that "in the implementation phase, the law could either be doomed to failure or improved and ambiguities clarified in order to make a modern social security service possible".

The reaction of the most important organization of the medical profession to the institution of the SSN is puzzling. On the one hand, it testifies to its limited influence on health policy, especially when compared with the bitter and protracted negotiations of the British Medical Association with the Minister of Health Aneurin Bevan in establishing the NHS (Marmor and Thomas 1972). In Italy, health reform had essentially been taken as part of the "democratic reform of the state" in implementing the Constitution. This transformed a health policy problem into a matter of "high politics", where FNOM appeared an unwelcomed and persistent guest. Moreover, the entry of the regions in the national political arena had broken the "iron triangle" between doctors, sickness funds, and ministerial bureaucracies that had been dominating for long the policy of the healthcare sector. Furthermore, processes of fragmentation and social and professional stratification that had been taking place within the medical profession had limited FNOM capacity to represent the cultural and professional interests of the whole profession.

On the other hand, FNOM's swift volte-face lent plausibility to adopting Rudolf Klein's interpretation of the battle between Aneurin Bevan and the British Medical Association on the creation of the British NHS (Klein 2006, p. 13). Klein argues that "it was in the political interests of everyone concerned to overstate the extent of disagreement". For Bevan, the fierce confrontation was a testimonial of his proposal's radicalism, while the BMA took it as an opportunity to conceal the compromises it was prepared to accept. The pragmatic attitude adopted by FNOM certainly reveals an acute tactical awareness of the complex relations between the passing of a law and its implementation, when principles normally give way to issues of opportunity and feasibility. This explains, for example, the patching of the greatest defeat of organized medicine, the institution of the *ruolo unico* (single or common role), encompassing all the health professions employed by the SSN. A common role denied doctors a special position and was a threat to medical dominance both in their relations with the other health professions and in contractual negotiations. The debate over the *ruolo unico* became a matter of political exchange for the commitment of the profession in the implementation of the reform. The compromise was achieved in steps, first establishing a separate area of negotiation in contracts for "matters of major medical importance", then through an agreement that was translated into law in 1987.

A Reform Out of Time

The health reform had serious problems almost from the day it was passed. Giovanni Berlinguer had warned that "the real health reform begins after the approval of this law" (Berlinguer 1979, p. 188), and its implementation would require "a strong political authority, a remarkable efficiency in the work of the government, a vast popular consensus" (p. 190). Years earlier, in presenting his bill, Minister Vittorino Colombo stressed the essential role that the regions would be called upon to play in the implementation of the reform (Colombo 1977). However, as the next paragraph will argue, this did not go smoothly. The central government's activism in implementing a strict budget policy and legislating ever-increasing users' co-payments for health services contrasted with the inertia in devising the first National Health Plan, which supposedly was to guide the implementation of the SSN. The regions immediately showed their political and administrative inadequacies, and the USL revealed the many weaknesses in the way they had been designed. Issues of more general import were however at stake with the operational failure of implementing the SSN.

The reform had apparently been approved in a relatively short time following a "big-bang" model (Klein 1995). It was instead the belated product of a thirty-year debate focused in the main over general principles, which was still seriously wanting in design and organization. This contributed to the peculiar combination of practical failure and ideal success that constitutes the fundamental characteristic of Law 833. Ambiguities of the law and shortcomings in its implementation notwithstanding, it was the abrupt change in the political and economic context that mostly hampered the SSN from its beginning. The reform that had not been passed in the years of the economic miracle had to be implemented in a period of stagnation and inflation resulting from the two oil crises of the 1970s, which put all the welfare systems of advanced countries in crisis. While other countries were adopting restrictive policies, Italy had passed a strongly expansive reform in healthcare, the sector chosen internationally as one of the main targets to contain public spending. Moreover, at the same time, the government decided to enter Italy into the European Monetary System, which required complying with external constraints, such as forbidding devaluation of the currency. The content of the law was also at odds with the emergent neoliberal ideology of the minimal state, signalled by the ascent to power of Ronald Reagan in the United States and Margaret Thatcher in England.

The institution of the SSN is a classic example of Peter Hall's third-order change, which simultaneously involved all three components of the policy including altering the settings of the policy instruments (first order change), replacing old instruments (second order), and then adopting new goals (Hall 1993). The SSN made entitlements universal and individual benefits uniform, or allegedly so. *Casse mutue* were replaced by USLs, responsible for the health of a geographically defined population, under municipal and regional authority. The passing of the law had primarily been an act of political willpower associated with factors endogenous to the political and institutional system of the moment, represented by the governments of national solidarity with the participation of the communist party and the strong push from below of social movements. The support for reform by trade unions and leftist parties is consistent with the classic analysis of the emergence of welfare systems in terms of "mobilization of the working classes" (Esping-Andersen 1990) and Korpi's theory of "power resources" (Korpi 1989). The greatest paradox in the history of the NHS, however, is that when it was established, the main political and social forces that had supported it for so long were in complete disarray. The social movements had already travelled a long stretch of their downward path, the so-called *riflusso* (ebb). The government of "national solidarity" that passed the law by a landslide majority had given way to the return to centrist governments, which entrusted the Ministry of Health to a member of the Liberal Party, the only party, along with the post-Fascists, that had voted against Law 833.

The Undoing of the Reform

Pressures to amend Law 833 began soon after it was passed with the dissolution of the heterogeneous coalition that had supported its enactment. It intensified as the USLs slowly began operation, which revealed the weaknesses in the way they had been designed, and as evidence accumulated of their wasteful and inefficient use of resources. Exogenous factors also hindered the implementation of the SSN. Two influential OECD Reports predicted an impending crisis of welfare systems based on three factors, which became a model for subsequent analyses. The first and paramount factor was the abrupt halt of economic growth caused by the two oil crises of the 1970s. A mounting fiscal revolt of the taxpayers came second, and the increase in the elderly population followed (OECD 1981, 1985). Recommendations called for selective social policies targeting those in "real" need, increasing productivity and eliminating waste, contrasting the power of professional groups, such as doctors and teachers, who had allegedly taken control of health and education, the two fastest growing sectors of the welfare state (Symposium 1985). Italy aligned with the international trend by adopting budget policies that in the health sector were associated with forms of selective or "conditional" universalism (Ferrera 1995). These were adopted through annual financial laws, which for the first half of the 1980s were the main instrument of governing the newborn SSN.

As evidences on USLs' mismanagement combined with the divide between central funding and local spending power, the conditions seemed to be in place for the cost of the SSN to grow fast. This did not occur as expected. The *Comitato nazionale per l'economia e il lavoro*, a national watchdog which, as we have seen, had little enthusiasm for the SSN, certified that medical spending as a share of GDP was essentially stable. Public health expenditure ratio to GNP stood before the reform at 5.6 per cent in 1976 and 1977 and 5.7 in 1978, rose to 6.0 in 1979, and then went down again to 5.8 per cent in both 1980 and 1981 (CNEL 1985). A series of decrees concerning "urgent financial provisions" froze service supply, such as opening new services and hiring personnel, and increased users' co-payments for drugs and medical services. The severity of the financial context in which the SSN was to be implemented reached its peak with the second finance law of April 1982 (Law nr. 526, August 7, 1982), which cut 4700 billion lire, or slightly less than 10 per cent, of the SSN budget. The law report explained that "for the first time since its establishment, the National Health Fund is not based on projected expenditure but depends on the financial resources which had been allocated to health". Its expected impact on the SSN was so disruptive that the report stated that "the fundamental principles of Law 833 remain valid and the reform process is not stopping", but it "was adapting to the changing and objective economic conditions of the country".

The policy of cost sharing developed along three lines. The scope of application of co-payment was expanded from drugs to specialist outpatient diagnostic and therapeutic services. The amount of co-payment was increased and took the form of a lump sum for medical and laboratory services or a proportion of drugs' selling prices, typically between 15 and 30 per cent, but up to 50 per cent for certain drugs. Finally, a ceiling on individual contributions per prescription was set, as well as a complex system of exemptions based on personal income, age, and type of disease (a detailed description is in Ferrera 1995). The government's financial activism contrasted with its inertia in devising the first National Health Plan, which was supposed to guide the implementation of the reform. The 1980–1982 plan, which would set standards of services, financial resources, and guidelines for implementing the USL, was due by September 1979, but it never came to light. After several drafts, the first National Health Plan was approved in 1994, fifteen years after the deadline and, ironically, associated with the "reform of the reform" (see next Chapter).

The regions had been one of the main actors of the institution of the SSN, and the enactment of the law marked their role as autonomous institutions in the national political arena. However, their role was weakened in favour of the municipalities that controlled the USLs and emerged as the operational centres of the new organizational structure. Moreover, regions' autonomy was compressed by the budget policies of the 1980s, which also put to test the limited fiscal capacity that the centralist tax reform of 1971 had granted them. In the main, the regions, especially in the south, gave little proof of themselves in implementing the SSN. A survey showed that by 1982 all regions, except Sicily, had formally set up the USLs, but the vast majority were still grappling with transferring them the assets of the hospitals and other health organizations (CNEL 1982). There was wide regional variation in the size of USLs, whose average population ranged from 44,000 inhabitants in Trentino and Molise to 178,000 in Friuli and 139,000 in Veneto and Lombardy compared to a national average of 87,000 inhabitants per USL (Del Vecchio 1980). Differences between regions in implementing the reform suggested the important conclusion that "the gap between the most advanced regions and those at the lowest level of development is increasing, an outcome which seriously contradicts one of the fundamental principles of the reform" (CNEL 1982, p. 98).

An additional important problem stemmed from the governance of the USL, with the responsibility for managing services and, at the same time, promoting democratic participation through political parties' appointees. Maurizio Ferrera (1989, p. 120) observed that in no other country "have the distinct tasks of democratic control and top management been entirely subsumed into a single elective organ exposed to such heavy and thoroughgoing party manipulation". The political class grasped the decentralization of powers created by the reform as "an unexpected and extraordinary opportunity to allocate resources, to channel them according to the usual particularistic and clientelist practices" (Fargion 1997, p. 124). In the name of participation, USLs were crammed with "young fixers on the rise, opaque second-row officials, apprentices ready to perform on new stages, ex-deputies, ex-mayors, ex-councilors and various retirees worthy of a place in their "small homelands" (Lanaro 1997, p. 362). The demand to "de-politicize" healthcare giving specific managerial responsibilities to technically qualified personnel was one of the main factors that, in Italy, supported the managerial revolution that had already established itself in the public administration of several countries.

A collection of practices commonly understood under the generic term of New Public Management (NPM) emerged in the 1980s and 1990s as the hegemonic paradigm in managing public administration. It aimed at increasing productivity and efficiency of public services by making them more akin to private enterprises and the competitive markets in which they traditionally operate. The great variety of forms assumed by the NPM in different countries has proved resistant to codification according to a precise model, so much so that it has been dubbed "a tool for all seasons" (Hood 1991). In Italy the term New Public Management has never been used in official documents, but its concepts have influenced the process of public administration reform, including healthcare. The promises of de-politicization and de-bureaucratization of service management were particularly attractive, as well as the search for greater attention to the preferences of users of the services, or "customers". The appeal to modernization and moralization combined with the seductive image of a management entrusted to responsible and competent technicians was perceived as a "moral antidote" to corruption and a protective shield for intrusiveness and the incompetence of politics.

While regionalization emerged later, the corporatization of the USLs was a hot topic of the debate of the late 1980s. The medical elites participated in the health policy debate with the *Manifesto bianco* (White Manifesto), signed in 1985 by seven famous hospital super-specialists, two of whom later became Ministers of Health with the centre-right and the centre-left coalition. The "White Manifesto" denounced the debasement of hospital medicine, resulting from its incorporation into the USLs. The priority given to prevention and community services had caused the decline of the largest hospitals where super-specialties were concentrated, which hindered the quality of care in "centers of excellence". The call was for a "special legislation" to take the big hospitals out of the USLs and make them autonomous organizations. They would be managed by an autonomous Board of Directors "free to manage in the ways it deems most appropriate to achieve efficiency goals" (Boeri et al. 1985; www.libertamedica.it/white.html).The proposal to amend Law 833 opened the way to a frenzied debate, and various governments began considering a "reform of the reform". The scope broadened to include, according to the bill of the Health Minister Francesco De Lorenzo, "elements of privatization and liberalization in the government, structure and functioning of the SSN". Various bills on these issues were shuttled between Chamber and Senate, and this was dubbed by Beniamino Andreatta, Treasury Minister in several governments of the time, the "goose game" of the reform of the 1978 healthcare reform (ISIS 1991). Eventually, two financial and judiciary shocks opened a window of opportunity to stop the game and pass the reform of the SSN reform.

References

Arbasino, A. (1980) Un paese senza. Milano, Garzanti

Babini, P. (2009) Liberi tutti. Bologna, Il Mulino

Barca, S. (2012) Bread and poison: stories of labor environmentalism in Italy, 1968–1998. In Sellers, C., Malling, J. Eds History of industrial hazards across a globalized work. Philadelphia, Temple University Press (pp. 126–139)

Basaglia, F. (1968) L'istituzione negata: Rapporto da un ospedale psichiatrico. Torino, Einaudi

Berlinguer, G. (1976) Malaria urbana. Milano, Feltrinelli

Berlinguer, G. (1979) Una riforma per la salute: Iter e obiettivi del Servizio sanitario nazionale. Bari, De Donato

Berlinguer, G., Biocca, M. (1987) Recent developments in occupational health in Italy. International Journal of Health Services 17 (3), pp. 455–474

Boston Women's Health Book Collective (1971) Our bodies, Ourselves. New York, Simon & Schuster

Brenna, A. (1974) L'economia del sistema sanitario italiano. Le Scienze, 73, pp. 110–122

Brown, L.D. (1984) Health reform, Italian style. Health Affairs, 3 (3), pp. 75–101

Buglione, E., France, G. (1983) Skewed fiscal federalism in Italy: implications for public expenditure control. Public Budgeting and Finance, 3 (3), pp. 43–63

Cavazzuti, F., Giannini, S. (1982) La riforma malata: un servizio sanitario da reinventare. Bologna, Il Mulino

Censis (1979) Rapporto sulla situazione sociale del paese, 1978. Roma

Centemeri, L. (2006) Ritorno a Seveso: Il danno ambientale, il suo riconoscimento, la sua riparazione. Milano, Mondadori

CNEL (1982) Osservazioni e proposte sullo stato di attuazione della riforma sanitaria. Roma n.90/138

CNEL (1985) Osservazioni e proposte sul finanziamento del SSN. Roma

Cochrane, A.L. (1972) Effectiveness and efficiency: Random reflections on health services. London, Nuffield Provincial Hospital Trust

Colombo, V. (1977) La giungla sanitaria: Idee per una riforma dei servizi sanitari. Torino, Vallecchi

Del Vecchio, G. (1980) Considerazioni sullo stato delle Unità sanitarie locali. Igiene e Sanità Pubblica, 36, pp. 48–57

Ehrenreich, B., Ehrenreich, J. (1971) The American Health Empire: Power, profits and politics. New York, Random House

Ergas, Y. (1982) 1968–79. Feminism and the Italian party system: Women's politics in a decade of turmoil. Comparative Politics, 14 (3), pp. 253–279

Ergas, Y. (1986) Nelle maglie della politica: Femminismo, istituzioni e politiche sociali nell'Italia degli anni '70. Milano, F. Angeli

Esping-Andersen, G. (1990) The three worlds of welfare capitalism. Cambridge, Polity Press

Fargion, V. (1997) Geografia della cittadinanza sociale in ItaliaBologna: Regioni e politiche assistenziali dagli anni settanta agli anni novanta. Bologna, Il Mulino

Ferrera, M. (1989) The politics of health reform: Origins and performance of the Italian health service in comparative perspective. In: Controlling medical professionals: The comparative performance of health governance. Freddi G., Bjorkman J.W., Eds. London, Sage, pp. 116–129

Ferrera, M. (1995) The rise and fall of democratic universalism: Health care reform in Italy, 1978–1994. Journal of Health Politics, Policy and Law, 20 (2), pp. 275–302

Fioritti, A. (2018) Is freedom (still) therapy? The 40th anniversary of the Italian mental health care reform. Epidemiology and Psychiatric Sciences, 27, pp. 319–323

Foot, J. (2015) The man who closed the asylums: Franco Basaglia and the revolution in mental health care. London, Verso

Freddi, G. (1990) Introduzione: cultura politica e professionalità medica. In: Freddi G. (a cura di) Medici e stato nel mondo occidentale. Cultura politica e professionalità medica. Bologna, Il Mulino, pp. 7–48

Ginsborg, P. (1990) A history of contemporary Italy: Society and Politics 1943–1988. London, Penguin Books

Golden, M.A. (2003) Electoral connections: The effects of the personal vote on political patronage, bureaucracy and legislation in postwar Italy. British Journal of Political Science, 33, pp. 189–212

Habermas, J. (1976) Legitimation crisis. London, Heinemann

Hall, P.A. (1993) Policy paradigms, social learning and the state: The case of economic policy-making in Britain. Comparative Politics, 25 (3), pp. 275–296
Hood, C. (1991) A public management for all seasons? Public Administration, 69, pp. 3–19
Illich, I. (1974) Limits to medicine: Medical nemesis: The expropriation of health. London, Calder and Boyars
ISIS (1991) Andreatta illustra la sua formula per indiretta e sgravi contributivi. ISIS, 27 maggio 1991, pp. 12–13
Jones, K., Poletti, A. (1985) Understanding the Italian experiment. British Journal of Psychiatry 146, pp. 341–347
Jones, K., Poletti, A. (1986) The Italian experience reconsidered. British Journal of Psychiatry 148, pp. 144–150
Klein, R. (1995) Big bang health care reform: Does it work? The case of Britain's 1991 National Health Service reform. Milbank Quarterly, 73 (3), pp. 299–337
Klein, R. (2006) The new politics of the NHS: From creation to reinvention. Oxford, Radcliffe Publishing
Korpi, W. (1989) Power, politics and state autonomy in the development of social citizenship. American Sociological Review, 54, pp. 309–328
Lalonde, M. (1974) A new perspective on the health of Canadians. A working document. Ottawa, Government of Canada
Lanaro, S. (1997) Storia dell'Italia Repubblicana. Venezia, Marsilio
Maccacaro, G.A. (1979) Per una medicina da rinnovare. Scritti 1966–1976. Milano, Feltrinelli
Marmor, T.R., Thomas, D. (1972) Doctors, politics and pay disputes: "Pressure Groups Politics" revisited. British Journal of Political Science, 2 (4), pp. 421–442
McKeown, T. (1976) The role of medicine: Dream, mirage or nemesis? London, Nuffield Provincial Hospital Trust
O'Connor, J. (1973) The fiscal crisis of the state. New York, St. Martin's Press
Oddone, I., Marri, G. (1977) Ambiente di lavoro. La fabbrica nel territorio. Roma, Editrice Sindacale Italiana
OECD, Organization for Economic Cooperation and Development (1981) The welfare state in crisis. Paris
OECD, Organization for Economic Cooperation and Development (1985) Social expenditure 1960–1990: Problems of growth and control. Paris
Pierson, P. (1994) Dismantling the Welfare State? Reagan, Thatcher and the politics of retrenchment. Cambridge, Cambridge University Press
Pierson, P. (1996) The new politics of the welfare state. World Politics, 48, pp. 143–179
Putnam, R.D., Leonardi, R., Nanetti, R.Y. (1983) Explaining institutional success: the case of Italian regional government. American Political Science Review, 77, pp. 55–74

Reich, M., Goldman, R. (1984) Italian occupational health: Concepts, conflicts, implications. American Journal of Public Health, 74 (9), pp. 1031–1041

Rosaia, L. (1976) (a cura di) L'ombrello bucato. Introduzione alla riforma sanitaria. Roma, Edizioni della Voce.

Spaventa, L. (1988) Introduction. In: Giavazzi F., Spaventa L., Eds. High public debt. The Italian experience. Cambridge, Cambridge University Press

Symposium (1985) Social expenditure: 1960–1990: Problems of growth and control. Journal of Public Policy, 5 (2), pp. 133–168

Taroni, F. (2011) Politiche sanitarie in Italia: Il futuro del Ssn in una prospettiva storica. Roma, Il Pensiero Scientifico

Taroni, F. (2015) Salute, sanità e regioni in un Servizio sanitario nazionale. In: Salvati M., Sciolla L., Eds. L'Italia e le sue Regioni. L'Età repubblicana. I, Roma, Edizioni Enciclopedia Treccani, pp. 411–427

CHAPTER 8

Reinventing the SSN?

In the 1990s, "a series of seismic shocks…radically transformed Italy's political landscape" (Rhodes 2015, p. 309). Of the two main shocks, the first was the discovery of a wide network of political corruption (*tangentopoli*—city of bribes) by the judiciary operation known as *mani pulite* (clean hands), which caused the crisis of political parties. The other was the *mercoledì nero* (Black Wednesday) of the Italian currency when the lira plummeted and dropped out of the European Monetary System. This fostered change in monetary, economic, and social policies, including healthcare. In the second half of the 1990s, political institutions also changed. The reconfiguration of the political party system and a new (partly) majoritarian electoral system made party politics less collusive and more adversarial. This produced first a short-lived centre-right government, followed by a series of technical governments, and then a centre-left government that ran its full term from 1996 to 2001. This was exceptional continuity for a government in Italy. Later in this period, the presidents of the regional governments were directly elected by popular vote, making the regions more assertive vis-à-vis the central government.

External shocks and institutional innovations introduced a new dynamic into Italian policymaking and significantly changed its form, process, and content. Comprehensive reforms were introduced in four major social policy areas (healthcare, social services, education, and labour), while a sweeping reorganization of the public administration was performed. Legislating on politically sensitive and technical complex sectors was the

F. Taroni, *Health and Healthcare Policy in Italy since 1861*,
https://doi.org/10.1007/978-3-030-88731-5_8

result of learning processes both from abroad and from experience, stimulated by the "enabling constraint" imposed by the Europeanization of the economic policy. After the blockade of healthcare policy of the 1980s, the new style of policymaking affected particularly the healthcare sector, which anticipated the emergence of many institutional innovations in public administration. Over the decade, there have been two distinct and opposing waves of reform of the SSN, characterized first by corporatization of the USLs, regionalization, and, in some regions, structural separation between purchasers and providers of health services, and then by planning, integration, and cooperation. Both waves followed the changing international context of healthcare reforms, and particularly the reforms and counter-reform of the British NHS. In addition, there was an important series of measures imposing curtailments in access to covered services and an experiment in "rational rationing" of drug coverage, including pruning the SSN pharmaceutical formulary based on clinical effectiveness and cost containment, a national drug budget, and new arrangements for setting drug prices and users' co-payment.

The Origins of the "reform of the 1978 reform"

As we have seen, pressures to amend Law 833 intensified as evidences emerged on the mismanagement of the USLs and their wasteful use of resources. Between 1980 and 1990, the annual expenditure overshooting of the SSN averaged about 10 per cent of the original spending ceiling, and its aggregate deficits totalled about thirty billion liras over a budget of about sixty billion liras per year. Various government bills proposed a reform of the 1978 reform, but it took two exogenous "seismic shocks" to break the legislative blockade.

The judiciary shock of *mani pulite* caused the acute crisis of the party political system. Several of the leading politicians of the ruling parties were charged with, and later found guilty of, collecting kickbacks to raise party funds and/or to benefit personally (McCarthy 1995). This led to the virtual disappearance of the main parties and the disbandment of their local constituencies. The other shock had two aspects to it. A fiscal crisis resulted from the emergency in the public finances caused by the "Black Wednesday" of the Italian currency in September 1992, when the lira went into a tailspin and dropped out of the European Monetary System. Public finances reached a double-digit budget deficit to GDP ratio. Expenditure cuts equivalent to 6 per cent of GDP were applied, and the health budget diminished in real terms for three consecutive years between 1993 and 1995. The second shock consisted in the external constraint deriving from the convergence criteria of the Maastricht Treaty for the Italian membership in the European Monetary Union. The Treaty defined a strict timetable for the meeting of budgetary targets, closely monitored by peer states, which were particularly demanding on Italy, which however took the challenge. Some scholars claimed that Europe had in fact "saved" or at least "rescued" Italy (Ferrera and Gualmini 2004). The article in the newspaper *La Repubblica* cited by Della Sala exemplifies the persuasive force of the appeal to the European constraint. Since many areas in Southern Italy were short of drinking water and other utilities, it was proposed to include the availability of drinking water, gas, and electricity among the criteria for entering the EU (Della Sala 1997). The external constraint of the EU was the strategic asset that technocratic cabinets used to help pushing through policies that otherwise could have remained bogged down. Guido Carli, former governor of the Bank of Italy, Minister of the Treasury, and member of the Italian delegation, explained the politics of "the two levels game" (Gourevitch 1978) in negotiating the Maastricht Treaty: "Our action at the table of the intergovernmental conference for the European Union was an alternative solution to domestic problems that we were not able to deal with the ordinary procedures of the government and the Parliament" (Carli 1993, p. 435).

The judiciary and financial watershed impressed a new dynamics in policymaking aimed at fostering change in monetary, economic, and social policies, where it focused on the four structural sources of public deficits comprising pensions, local government finance, public administration, and healthcare. The critical juncture of the crises opened a window of opportunity for the legislative reform of the SSN, which had been impossible to pass in the 1980s. The reform took most of the ideas that had been floating around for many years, waiting for the right political event to be enacted, a paradigmatic example of Kingdon's model of policy change (Kingdon 1994). Both the critical juncture, which allowed overcoming the legislative blockage, and the political agenda, which characterized that window, went beyond the health policymaking domain and concerned wider domestic policies as well as the process of Europeanization. To prevent Italy becoming the "Disneyland of Europe", as the President of the Council of Ministers Giuliano Amato famously put it in his acceptance speech in Parliament (see, e.g., *L'Unità* and *The Times*, July 1, 1992), the government developed a long-term political and institutional agenda to "modernize the country" with a package of both macro- and micro-economic reforms. The strategy was threefold: hard budgetary politics combining tax increase and spending cuts; a wide-reaching plan for privatization of state-owned enterprises; social reforms of pensions, health services, public administration, and local government finance. The modernization agenda explicitly aimed at improving the competitiveness of the state in the world economy by streamlining social policies based on market principles. In this agenda, the long-term goal of reforming the SSN loomed less than its immediate contribution to curb public expenditure.

The Content: An Assemblage

The 1992 health reform was an assemblage of many different streams centred on three core elements. First, a hard budget constraint was applied to the national health budget that was defined based on the resources allocated by the Treasury, not from projected expenditure. The new official policy was that uniformity and comprehensiveness of the services delivered by the SSN were not citizens' entitlements but represented "an objective to be pursued insofar as the available resources allow". Second, budgetary and organizational responsibility over health services was devolved down to the regions (the so-called *regionalizzazione* or regionalization). The official goal was "to put the decisions over the organization of services closer and more responsive to the needs of the people", but this also shifted the responsibility for overshooting regional budgets allocated by the central government to the regions themselves, out of their own resources. To compensate for their inadequate fiscal capacity, regions were allowed to let part of their citizens to opt out of the SSN in favour of private insurance plans or participate in public-private initiatives for concessions and franchising of public facilities. The "exit" of part of the population from public coverage was the only possible response to declining central transfers and inadequate own-sources revenues. Institutionally, regionalization sidelined the municipalities from running of the SSN. This gave the regions still greater power in the administration and organization of health services in exchange for their accepting harder budget constraints, but seriously hampered the integration between health and social services, the preserve of municipalities. Third, USLs were to be transformed into semi-autonomous public enterprises (*aziende sanitarie locali—ASL* or Healthcare enterprises). Legally, ASL were public agencies created by the region and run by *Direttori generali* (general managers), seeking efficiency in the production and delivery of health services by borrowing managerial tools and methods from the private sector (the so-called *aziendalizzazione* or corporatization). In addition, the central government could authorize the regions to hive major hospitals off *aziende sanitarie locali* and transform tertiary care and teaching hospitals into independent *aziende ospedaliere* (Hospital Enterprises), (very) broadly similar to British hospital trusts.

The *aziendalizzazione* or corporatization of USLs is a fundamental event in the history of the SSN, loaded with practical and symbolic meanings. Their constitution in autonomous health agencies ended the long debate on the nature of the USL and transformed the "mysterious object" of the 1978 debate into a hybrid public agency combining public ownership and financing with entrepreneurial autonomy "based on criteria of efficiency, cost-effectiveness and effectiveness". ASLs were however subject to administrative law in procurement and hiring, and not allowed to keep annual income. The increase of the size of the ASLs (also a problem with Law 833 in 1978) from 87,000 to 290,000 inhabitants on average drastically reduced their number from 659 to 228. This indirectly further reduced the influence of the municipalities, which had been sidelined from their government in favour of the regions. The *direttore generale* was the most visible innovation and indeed became the symbol of the reform itself. The creation of the post and the new regulation of hiring and paying according to private rules applied to the SSN the principles of the reform of the public administration, which was proceeding at the same time and was part of the same reform agenda (Capano 2003). The new model of technocratic politics met the expectations of freeing healthcare from the grip of politicians, although the *direttori generali* were also political appointees of the president of the region, under the thin veil of a national register to certify managerial experience, later abolished and now reinstated. On the other hand, the broad powers of *direttori generali* broke the traditional, if mostly implicit, system of consensus management with doctors, which ensured them broad autonomy over clinical decisions and, to some extent, organization of practice. The assumption was that a system of "high-powered" incentives would guide clinicians' behaviour and streamline production processes.

Government's managerial strategies were sketched out only in broad-brush terms, supposedly out of respect for the autonomy of the regions that were left to fill in the details of the "bright ideas" of the law and possibly following the principles of "steering not rowing" and "letting managers manage" (Osborne and Gaebler 1992). Its emphasis on *aziendalizzazione* and managerialism refers more to the 1983 Griffiths' NHS Management Inquiry than to the 1989 White paper "Working for patients" (Gorsky 2013). Although the law was weak on details, it seemed however to offer the opportunity for de-integrating the highly vertically integrated SSN into a sort of British internal or quasi-market (Le Grand 1991). In Italy, however, the separation of health service providers from their "purchasers" (the so-called purchaser-providers split upon which the British internal market was based) crucially depended on how many hospitals the region proposed and the ministry authorized to constitute in independent *aziende ospedaliere,* and the adoption of market-based mechanisms by the SSN was not explicitly envisaged.

The regionalization of policymaking and the application of the enterprise strategy to health service production and delivery were purported as the new managerial instruments for the pursuit of the original goals of the SSN, which apparently were not directly challenged. The 1992 reform is therefore to be seen, in Peter Hall's terminology, as enacting first- and second-order change, given its formal adherence to the objectives contained in the 1978 reform law (Hall 1993). However, its long-term effects were more uncertain. Some clearly pro-market provisions, such as the "exit" of part of the population towards "alternative providers" and the incentives to Private finance initiatives or Public-private partnerships, were widely perceived to put the SSN down a slippery slope towards creeping privatization of the public system, as it was happening with the British NHS (Pollock 2004). Moreover, a "two-tiered" health system had been explicitly envisaged a few month before the reform was passed. The Law Decree nr. 384 of September 19, 1992, containing "Urgent measures in the field of social security, health and public employment, as well as tax provisions" provided for "citizens belonging to households with a total income in 1991 exceeding forty-one million lire [...] the cessation of basic medical assistance, of pharmaceutical assistance, with the exclusion of life-saving drugs, of instrumental and laboratory diagnostics and other specialist services, including rehabilitation as well as spa treatments" (art. 6, para.1). The household income threshold excluded from SSN coverage, except hospitalization, an estimated twenty-four million people or about 40 per cent of the Italian population. To the objection that families with an yearly income of forty-one million could hardly be considered well-off Prime Minister Amato's response was frank and explicit "No. However, we must start from a threshold. We have chosen the highest threshold which allows the final result" (La Repubblica, September 27 1992, pg.1). The provision was dropped when the decree was converted into law and emerged later in a softer version as an opportunity for the region in difficulty with their allotted budget.

A Swift Process

The 1992 reform was but one of a vast and ambitious agenda for social reform adopted by a government without a direct electoral mandate in a situation of political and financial emergency. Essentially its mandate was defined using an exceptional procedure, a *legge delega* (delegatory law) followed by a *decreto legislativo* (legislative decree).The delegatory law was approved by the Parliament and authorized the executive to draft a legislative decree, along the provisions broadly defined by the Parliament. The decree was subsequently subject to parliamentary scrutiny only for its conformity with the general indications contained in the delegatory law (Mattei 2007). This procedure meant that the government had control over lawmaking activity (in ordinary circumstances the exclusive province of Parliament in Italy), thereby avoiding lengthy parliamentary negotiation for enacting contentious structural reforms in sensitive social areas. Parliament's fragility was demonstrated by the fact that judicial investigations lead to one-third of the members of the main chamber, the Chamber of Deputies, being under investigation or facing corruption charges (McCarthy 1995). In a context of political and institutional crisis, in which political parties and the Parliament itself had been swept up by *mani pulite*, the use of delegated legislation hardened government's capacity to steer the parliamentary process, something which had been notoriously deficient up until then (Fabbrini 2000). The process clearly emerges from President Amato's vivid description: "Aware of the growing difficulty of obtaining a collective consensus via party political and parliamentary channels, I turned to the organisations representing industry and trade and the trade unions as alternative channels which had, in that period, a more direct contact with public opinion [...]; I presented the bill in Parliament and requested a vote of confidence on it, empowered by the knowledge that I had that consensus obtained outside Parliament" (Amato 1994, p. 366).

The "neo-corporatist" bargain (Fabbrini 2000) exposed the role of a small group of technocrats as the "policy entrepreneurs" in introducing social reforms. The vast reform package approved in the early 1990s has been described as an elite project for modernizing the Italian state. It initiated with the fiscal crisis and under the imperative of the *vincolo esterno* (external constraint) imposed by the European treaty and was designed by a small group of technocrats that shared common concerns about problems of governance and pursued similar solutions across the different fields of monetary, economic, and social policy (Dyson and Featherstone 1996). A series of short-lasting technical governments of experts from the Bank of Italy, private banks, public and private industry, and the universities interlocked and ensured stability in focus and content throughout the 1990s. Scholars observed that the exceptional character of the moment overturned the paradox of the so-called first Italian republic: governments which survived without governing were replaced by executives "governing without surviving" (Capano and Giuliani 2001). The transnational network of which Italian technocrats were part helped the transnational transfer of economic and financial policies (Quaglia 2005). It is less clear that similar transfers happened also with the health reform. Transnational learning in the healthcare field was common, particularly from the United States, about both broad policy issues, such as managed care (Marmor and Plowden 1991), and specific tools, such as patient classification systems for paying the hospital (Fetter et al. 1980) and quality improvement (Gonnella et al. 1984). In Britain, the debate over the health reform was largely shaped by Einthoven's influential lecture on the management of the NHS (Enthoven 1985). Working up the 1992 reform entailed quite a different process from the original 1978 legislation. Evidence is that the 1992 reform was drafted in haste without much consultation even of the most significant stakeholders, such as regions and medical organizations (France and Taroni 2005). In this it bears some similarities with Margaret Thatcher's White paper "Working for patients" in England (Timmins 1996, p. 453). A translation of the 1989 UK White Paper "Working for patients" with a companion volume discussing the potential transfer to Italy (France 1990) was made available by George France and the *Istituto di studio delle Regioni* (Institute for Regional Studies) of the National Research Council (Istituto di Studio delle Regioni 1990), along with a collection of papers on the Scandinavian model of "public competition" (Saltman and von Otter 1991). However, no documentary evidence can be found of "social learning" actually occurring in drafting the 1992 law. The introduction of competition and managerialism in the SSN seems therefore less the result of a process of transnational learning than the ripple effect of the general strategy for reforming public sector administration, which found its first implementation in the health field.

Implementation and Its Impact

A politically sensitive and technically complex health reform was enacted in the record time of a few months, but its implementation was staggered over time. This was in part the traditional problem of the inertia of the Italian public administration, which was now complicated by the new responsibilities devolved to the regions. The first regional laws instituting the new *Aziende sanitarie*, for example, had to wait until 1995; hospital payment by Diagnosis-related groups (DRGs) was officially introduced in 1994 (Taroni 1996), and it took several years before it was regionally implemented (Falcitelli and Langiano 2004). Italy seemed bent on joining the quasi-market movement of healthcare reform, but the development of some form of competition in healthcare was hard to believe. For example, the geographical distribution of hospital specialties revealed, for example, widespread natural monopoly and monopsony to allow for the functioning of an internal market of a sort. The chief executive officers of the USLs and hospitals, the leading actors in the new healthcare system, had very imprecise ideas about competition in general and in the health sector in particular, as demonstrated by in-depth interviews (France 1999). Moreover, hiving off public hospitals was never intended to be total, but rather to be limited to hospitals of "national importance" and did not include out-of-hospital specialist care. Furthermore, ASLs had little possibility of acting as "purchasers" because the region was the dominant player in setting up and managing the whole system. Most regions financed *aziende ospedaliere* directly and negotiated collective agreements with private contracted hospitals, amounting to around 20 per cent of total national capacity. Other regions instead formally transferred the bulk of available resources to ASLs, but ring-fenced the funds the region decided to allocate to *aziende ospedaliere*, or set individual spending caps.

The Amato government was followed by a series of short-lived technical governments, which kept tinkering with the original law, although their interlocking contributed to maintain stability in focus and content. At first, Azeglio Ciampi's government revised the most market-oriented parts of the 1992 law that had been threatened of abrogation through a popular referendum by the trade unions and the regions. These included "alternative" private insurance plans for citizens opting out of the SSN and schemes for concessions and franchising of public facilities similar to the British Private Finance Initiative. The new Health Minister Maria Pia Garavaglia transformed the "alternative forms of assistance" provided for by art. 10 of Legislative Decree 502/92 in "Supplementary forms aimed at providing additional services to those insured by the National Health Service" with Legislative Decree no. 517 approved in December 1993. Then, a series of Budget Acts further regulated the relations between the regions, healthcare enterprises, and private providers of which the 1992 law had disposed with under the vague qualification of *appositi rapporti* (special relations). Filling in the gaps was essentially a process of learning by doing which implied developing a system of accreditation for public and private providers, a new DRG-based system of hospital payment and the role of the contractual agreements between regions, *aziende sanitarie*, and private, accredited providers. This caused a "permanent revolution" in the relations between the main actors of the SSN, including the state and the regions, the regions with each other and with their *aziende sanitarie locali*, as well as between them, private providers, and *aziende ospedaliere*. Frequent changes in the complex bundle of regulations offered the empowered regions various choices over their model of organization, according to their administrative capacity, political stability, and ideological orientation. The origin of most of the differences between regional health services is probably from this period, although it is generally ascribed to federalist instances which appeared only later.

This also explains in part why in Italy an internal market never really materialized. While in Britain the central government could reasonably aspire to introduce a standard model of healthcare organization throughout the country, in Italy the national legislation was essentially interpreted as a general framework. As responsibility for organization lay with the regions well before it was officially sanctioned with the 2001 Constitutional reform, they implemented those elements of the national framework that best met their administrative capability, political will, and ideological preferences. Particularly after the presidents of the regions were directly elected by popular vote, the organization of healthcare became a powerful element for asserting the political identity of the region, vis-à-vis the state and the other regions, as with the long-standing contrast between the competitive and pro-choice model of Lombardy and the more planned and cooperative model of Emilia-Romagna. Public health spending fell in real terms in 1993, 1994, and 1995, when its share on GDP went as low as 5.3 per cent. However, the impact of the 1992 reform on health expenditure was modest. The massive cut of public health expenditure was mostly provided by hardening the traditional policies of underfunding, service curtailments, and increased co-payment, particularly on drugs.

"Rational rationing" of Drug Benefits

In the early 1990s, aggressive pharmaceutical policies laid out in few months an unprecedented regulatory model both in process and in results, if not in content. The revolution wrought on the three components of drug policy, that is, drug selection, drug prices, and levels of users' co-payment, is an exemplar of explicit rationing on a scientific basis by a technical body with a specific political mandate to contain public spending within a predetermined ceiling. The main actors were a small group of highly regarded experts from outside the public administration appointed to a *Commissione Unica del Farmaco* (CUF-single pharmaceutical commission). The Commission was given free reign over a field traditionally reserved to opaque negotiations between powerful interest groups and various ministries because a judicial crisis had upset the whole old regulatory machine. The Health Minister Francesco De Lorenzo, the director of the pharmaceutical service Danilo Poggiolini, the president and numerous members of the Interministerial Drug Pricing Committee, the president of the association of the pharmaceutical industries, and several of its industrial associates had been arrested and subsequently sentenced to jail or pecuniary sentences (Minerva 2009, pp. 293–296).

The Commission's task was to select the drugs that the SSN would cover according to a new National Therapeutic Formulary; identify new criteria for objectively fixing the price of drugs, based on a European Average Price; and determine the amount of users' cost sharing in order to keep public pharmaceutical spending below the annual ceiling set by the government. Within three months, the CUF reclassified all drugs on the market into three classes. Class A included "essential drugs and drugs for chronic diseases", which were fully covered by the SSN. Class B contained "drugs of significant therapeutic interest", of proven efficacy but copies ("me-too" drugs) or combinations of drugs included in class A. They were subject to a 50 per cent cost sharing. Class C was a default category for all other drugs, including those whose price did not conform to the European Average Price. Drugs in class C were charged to the patient.

The net effect of the new classification was a sharp reduction in the proportion of drugs covered by the SSN over those in the market, which went from 62.3 per cent in 1993 to 54 per cent in 1994 and fell to 46 per cent in 1997 (Fattore and Jommi 1998). Between 1992 and 1994, public drug expenditure fell by 40.5 per cent, from 13.3 to 11.0 per cent of total health spending. Revenues from patients' co-payment from class A and class B drugs were reduced, but private spending on drugs increased, despite the marked drop in the consumption of some of the best-selling drugs that had been included in class C (Fattore and Jommi 1998). Apparently, a significant part of public expenditure had been shifted to the patients. Despite a general change in the "culture" of drug use (Garattini 1995), doctors' perseverance in their prescribing behaviours shifted to their patients the cost of rationalizing and moralizing the whole process of drug regulation, at least in part.

Such projects had not been unprecedented in the history of Italian healthcare. The *piccola riforma* of INAM in 1958 included a commission for a new drug formulary that ended up actually increasing the number of medical specialties covered. In 1970, INAM appointed a consultative Commission with the explicit goal of pruning its formulary. The result however had not been endorsed by the INAM Board on account of the fear that this would result in transferring to patients the cost of the discarded drugs (Scarpa and Chiti 1975, p. 171). In the new era, the pharmaceutical industry has been weakened by the scandals, which had also relegated politicians' interests to the sideline. "Experts were given pride of place and armed with regulatory powers with teeth" (France and Taroni 2005, p. 178), and the results were accepted by citizens as a minor trouble of fighting corruption.

Back to Base: The 1999 Reform of the 1992 Counter-reform

The 1999 reform, usually named after Rosy Bindi, the Minister of Health who made it pass, was an act of political will on the part of the centre-left government, which won the 1996 general election over the centre-right coalition of Silvio Berlusconi. A reform of the healthcare system was not initially part of the programme of the centre-left coalition, and health had not been a significant electoral issue (a relevant difference with the United Kingdom, where the Labour party had campaigned widely against the

Thatcher's health reform). Moreover, the health chapter of the government's Commission for the analysis of macroeconomic compatibility of social spending claimed continuity with previous policies (Onofri 2008). The Report claimed European health policies converging towards the model of "quasi-markets" or "internal markets" and considered "selective" universalism in terms of beneficiaries and/or type of services. Popular perceptions were somewhat different and became relevant in the new partisan political environment. The 1992 reform was alleged to encourage a narrow cost-paring mentality and a democratic deficit, limiting the voice of doctors and local authorities and giving too much power to the general managers. In addition, significant political choices were on the table. The vagueness and ambiguity of the 1992 law had encouraged opportunistic interpretations by some of the newly empowered regions. Clashes had occurred with the Ministry of Health, for example, when the region of Lombardy, run by a centre-right government, introduced an almost total separation between purchasing and provision. Confindustria, the national organization of Italian business, campaigned through position papers and conferences for a radical, Dekker-style reform of the Italian healthcare system where competing private insurance plans would provide patient choice and plurality of provision and funding (Confindustria 1997). According to influential scholars in 1997 it was "not wholly unthinkable" that "Italy would give up the universalistic principle of the SSN that it had first adopted among the countries of continental Europe" (France 1997). The situation was apparently similar to that observed in England when Kennet Clarke's comments on the White Paper admitted that "almost everything in the reforms, if not intended to lead to privatisation of the service, would unquestionably make any subsequent privatisation simpler" (Timmins 1996, p. 470). The intent of the Health Minister Rosy Bindi was "to do away with the anomalies of the counterreform of 1992" (Bindi 2005, p. 32) countering the "creeping privatization" of the SSN and the *federalismo da abbandono* (derelicting federalism) of the relations between the state and the regions. The goal of Bindi reform was to reactivate the fundamental principles of the SSN of universal, fair, and uniform access to a comprehensive package of health services and restore the public service ethos by remedying the effects which its opportunistic implementation was producing (Bindi 2005, p. 146).

The reform set three explicit goals. The first was to establish an essential and uniform benefit package (*livelli essenziali ed uniformi di assistenza*—LEA) to be uniformly provided by the SSN free at the point of consumption as a right of citizenship. The selection of LEA from what was currently offered by the SSN was essentially following the "sieve" concept of the Manning Commission of the Netherland based on criteria of clinical effectiveness, organizational and clinical appropriateness, human dignity, and cost-effectiveness (Taroni 2001). The second objective was to halt the process of creeping privatization and marketization of healthcare by limiting the constitution of *aziende ospedaliere*, setting tighter criteria for public-private partnerships and regulating the dual practice of doctors employed with the SSN. This implied setting national standards for the accreditation of services and regulating contractual agreements and the payment system for public and private providers. The third was to give priority to developing community services and primary care and to the integration of health and social care, to which previous initiatives had paid only lip service, concentrating on the reorganization of hospital care. On this, the law relied on the strategy of the National Health Plan approved the year before. The Plan, significatively named *Patto di solidarietà per la salute* (Solidarity Pact for health), set specific health targets to be pursued through the joint action of the SSN, municipalities, non-profit organizations, and citizens (France and Taroni 2000). This implied strengthening the participation of municipalities in the government of the SSN and the involvement of clinicians in the government of the *aziende sanitarie* adapting the so-called clinical governance that the British NHS was experimenting.

Bindi's health policy also began an incremental shift in the process of allocating the SSN budget to each region with the goal of redressing the wide inequalities in services' provision and access. The new needs-based population formula considered not just the size of the regional population but also its demographic and health profile. Expected usage of health resources was estimated according to the age and gender distribution of the regional population and national utilization rates of hospital, drugs, and specialty ambulatory services, adjusted by regional Standardized Mortality Ratios (SMRs) as a proxy for regional morbidity (France et al. 2005). Debates focused both on technical issues of the formula and on more political, distributional questions. The former included, for example, the proper weight for age as a need indicator, and on SMRs as proxy for morbidity. More political questions were raised by the southern regions, which asked for including measures of social deprivation to compensate for their younger demographic profile. The new re-distributive formula and the process through which central health funding was allocated to the regions however became established in health policy.

The 1999 reform is a paradigmatic example of Hall's first- and second-order change, where new tools and new settings were developed to pursue the fundamental principles of the SSN. There was considerable borrowing from the 1997 UK White Paper "The New NHS: Modern, Dependable" (Department of Health 1997) and its companion "A First Class Service: Quality in the New NHS" (Department of Health 1998). The 1997 UK White Paper was seen as an inspiration of a sort for designing a healthcare system characterized by cooperation rather than competition, the development of strategic partnerships with local governments, and the third sector in primary and community care and the development of clinical governance. The influence of the British White Papers is to be explained also by symbolic political reasons. The New NHS was perceived as a third-way-style political response to pro-market policies, which the Labour Party had campaigned against, and this sat well also with the political motives for the Italian reform. Policy transfer was intense and widespread to most of the actors involved in the debate. The new method to allocate funding to the regions more equitably was modelled on the British Resource Allocation Working Party (RAWP), and particularly its 1988 Review (Gorsky Millward 2018). The development of the new DRG-based Prospective Payment System for hospitals was participated by a dense network of international and regional experts, working in close contact with government officials (Taroni 1996; Falcitelli and Langiano 2004). The Lombardy region was broadly supported in its integral split by Italian and English experts, while Confindustria borrowed from the Netherlands its Dekker-style insurance reform, as the national government did for the process of selecting the basket of essential services. The impact of transnational experiences on the Italian reform indicates the growing influence of European models on civil servants and an end to the separation between civil servants and the transnational academic networks of which Italian researchers were part. The more intense interaction between ministerial bureaucrats and independent researchers was encouraged by the new rules regulating the recruitment of top government officials (a product of the 1993 reform of the public administration), which permitted hiring outwith traditional career structures.

A Turbulent Process

The 1999 reform proved to be politically divisive, even within the centre-left government coalition, both when getting it through Parliament and in its implementation. Enactment was secured by means of the expedient procedure of the *decreto legislativo*, the same used with the 1992 reform, which speeded up passage of the law but did not spare it from intense political scrutiny. Three issues were particularly controversial. The 1999 law was a comprehensive and relatively detailed blueprint with a rationalist view of policy design, and its heavy regulatory part had an obvious top-down approach. Both flew in the face of the conventional wisdom of the day. According to Sabino Cassese, a respected scholar of public administration and former Minister "the health service produced by the new legislative decree is an imperial health service, structured in the form of a pyramid and symmetrical, which may respond to the needs of the central government, but not necessarily to those of the collectivity" (Cassese 1999). During the 1990s pressures rose to opening the SSN to private sources of funding, on top of what the SSN was willing and able to provide, along the model already adopted with the pension reform (Cazzola 1997; Piperno 1997). Such ideas were particularly intriguing as they were frequently supported with extensive programmes of rationing public services that would easily transform into additional services part of what the SSN was currently providing. On this point, Bindi's reform was effective not so much in the fiscal regulation of *fondi integrativi* (supplementary insurance funds) as only contribution to supplementary funds was eligible for tax deduction, but also on defining the depth of the universal guarantees that the SSN ought to ensure through its "essential", as contrasted with a bare minimum, benefit package. The closure to multiple sources of funding however attracted harsh criticism as a second pillar, and in case a third one, was felt necessary for bridging what was described as "the gap between a constantly expanding demand and the declining capacity of public financing" (Amato 1999).

The most heated issue was the regulation of the vexed question of doctors' dual practice. Private medical practice within the SSN was based in the main on hospital pay beds to which doctors could admit fee-paying patients. This legally came from the 1938 Petragnani law, which implemented an administrative order issued in 1934 (see Chap. 4) and was a legacy of a much older tradition of a privilege accorded to *primari* by Opere pie. Pay or private beds were relatively few in numbers, but their symbolic and ideological significance was much greater. Pay beds introduced a double standard within the SSN, where some patients were treated according to their ability to pay instead of medical need. In practice, this allowed private patients to jump the queue for treatment, ahead of other patients in the waiting list. On the other hand, private patients provided a nice additional income to doctors in some specialties, such as surgery and orthopaedics, and others traditionally neglected by the SSN such as ophtalmology and dentistry. The clash of ideology and interest determined a fierce battle, similar to that engaged by the Secretary of State for Social Services Barbara Castles in 1974 (Klein 2006, p. 86). According to the first version of the law for doctors hired after enactment of the law, private practice could be allowed inside (the so-called *intra-moenia* or indoor) but outside the structures of the SSN was quite simply prohibited. Doctors already employed by the SSN had to choose, on a once-and-for-all basis, between undertaking their private practice within public facilities and outside. Those opting for the former had to share their fees to cover the operating cost of the *Aziende sanitarie,* keep the volume of private activity compatible with institutional practice, and have their waiting lists monitored to prevent inducing patients to choose going privately to jump the institutional waiting list. Those opting for the latter had their salary curtailed and were excluded from holding senior posts in the SSN. Further negotiations under the threat of a doctors' general strike that would have killed the law somewhat relaxed initial provisions and granted generous bonuses to doctors choosing to practise privately solely within the SSN. In addition, medical trade unions were given a direct say in the administration of the new regulations at the local level. This caused a fragmentation of the health policy arena, with the reconfiguration of a solid coalition between the National Federation of Medical Doctors (FNOM) and the main trade union for SSN doctors (ANAAO). Opposition faded except from doctors in specialties such as dentistry and eye clinics, where private practice was widespread. Appeals were made to the President of the Republic against the law in paid advertisements in the main newspapers. Eventually, over 90 per cent chose the first option, with significant variation between specialties (only 57 per cent of dentists chose, e.g., public practice) and across regions, where adhesion to public practice ranged from 81 to 97 per cent.

While the impact of the 1992 law was staggered over time, and ended in a variety of regional organizational models, the comprehensive and internally coherent Bindi reform was either implemented in full or rejected in its entirety. Some regions appealed the Constitutional Court, lamenting various limitations in their legitimate powers over the organization of health services (Balduzzi 2005). The centre-right coalition promised during the election campaign that it would dismantle the 1999 reform, but when it rose to power made small changes, including removing central government assent over creating *aziende ospedaliere* and trying out public-private partnerships in building or expanding hospitals. In the event, the regions failed to exploit their additional autonomy. In the year 2000, only ninety-four hospitals, with just under 28 per cent of the national public bed stock, had been hived off, and twenty-nine, or less than a third, of these hospitals were located in the Lombardy region (France et al. 2005). This again followed the international mainstream, as the fads in medical care management and policy of the 1980s and early 1990s were vanishing (Marmor 2004). The purchaser-provider split had been supplanted in Britain by cooperation and consolidation into vertically integrated health systems, through long-term and relational contracts (Light 1997; Goddard and Mannion 1998) and by mergers and integrations in the United States (Gaynor et al. 2015). In England, the Private Finance Initiative (PFI)—later called Public-Private Partnership—had attracted serious criticisms on the ground of poor value for money and the disruption of capital planning as well as for its scope for corruption and for opening the door to privatisation (Pollock 2004). The British Medical Association had campaigned against it, and was famously dubbed a "Perfidious Financial Idiocy" by an editorial on the *British Medical Journal* (Smith 1999). In Italy, when some of the few private finance initiatives entrusting public structures to private entities to compensate for the capital invested in their construction were assessed, their result was in line with "the most pessimistic conclusions of the English experience" (Fiorentini 2000). The limits on outside private practice were eventually lifted, but only a very small proportion of SSN doctors resumed external practice. Examining "policy in time" (Pierson 2004) is the essence for evaluating how complex processes unfold. If the observation window is open for too little time, we may be led to judge a reform inconsequential when on the contrary it produced substantive long-term effects. This is, for example, the case with the notions of *livelli essenziali di assistenza*, and clinical and organizational appropriateness introduced by the 1999 reform were initially neglected amid the heated debates and eventually became the main planks of healthcare policymaking up to now.

After the blockade of healthcare policy of the 1980s, the 1990s brought about an intense dynamism and a new style of health policymaking. State capacity (defined as its capacity for "learning the skills needed to translate its intentions into successful outcomes"—Skocpol and Finegold 1982) increased, and policy transfer was more frequent. The Europeanization of domestic policy also played an important role: the "logic of no alternative", along with a hearty dose of blame shifting, and national pride was instrumental in eliminating partisan considerations from the discussion of dramatic measures in domestic economic and social policies. All these factors combined in different ways in two major extensive reviews of the SSN. Managerialism and regionalization brought about by the 1992 reform were responses both to external judiciary and fiscal shocks and to the perceived failure of the 1978 reform that had encouraged soft budget constraints in the regions and waste, political patronage, and corruption in the USLs. The 1999 reform, in contrast, had its origins in the belief that previous changes in the way of competition and privatization had gone too far and were compromising the fundamental principles of the SSN.

How health policy was developed, its contents and where it took place also changed. By the end of the decade, the regions were powerful veto points in the formulation of national policies, and policy was made less in Parliament and in the central ministries and more through negotiation between the national government and the regions, something akin to "executive" or "cabinet federalism" (Watts 1989) familiar to Canadians. The regions have had time to mature as institutions, and their capacity for social learning and policy transfer (including between themselves) had improved, but the extent to which this occurred varied widely. At the end of the 1990s, organizational issues were no longer at the centre of the healthcare policy debate. The organization and administration of health services were generally accepted as a regional responsibility even before this was written in the constitutional reform of 2001 (see page 230). This changed the focus of the national health policy debate that moved to the political and fiscal dimensions of devolution. A major downward shift of responsibility for organizing and operating health services would ideally be accompanied by devolution of responsibility for financing. Instead, own-source revenues over which regions have autonomy remained a negligible part of their budget, purporting a new form of the traditional divide between spending power and public financing. On the other hand, any significant attempt to move towards some form of fiscal federalism was constrained by the fiscal imperative to contain public expenditure and to meet the public budgetary and debt standards set by the European Monetary Union, under the shadow of a towering public debt, a structural factor of Italian policy. This asymmetry between the central political and fiscal planks of devolution made the intergovernmental relations between the state and the regions and among the regions themselves to be conflict-ridden and gave way to a renewed governmental strategy of underfinancing and overspending. The new politics of healthcare could therefore further increase the fragmentation of the SSN and worsen its regional inequalities in access and outcomes. The new century opened up with a new version of déjà vu all over again.

References

Amato, G. (1994) Un governo nella transizione: la mia esperienza di Presidente del Consiglio. Quaderni Costituzionali, 14 (3), pp. 355–371

Amato, G. (1999) Tavola rotonda. In: Amato, G., Cassese S., Turchetti, G., Varaldo, R., Eds Il governo della sanità. Milano, F. Angeli, pp. 123–128

Balduzzi, R. (2005) Esiste ancora un Servizio sanitario nazionale? In: AA.VV. Il governo della salute. Regionalismo e diritti di cittadinanza. Roma, Formez, pp. 23–34

Bindi, R. (2005) La salute impaziente. Milano, Jaka Book

Capano, G. (2003) Administrative traditions and policy change: when policy paradigms matter. The case of Italian administrative reform during the 1990s. Public Administration, 81 (4), pp. 781–801

Capano, G., Giuliani, P. (2001) Governing without surviving? An Italian paradox. Law making in Italy, 1987–2001. Journal of Legislative Studies, 7 (4), pp. 13–36

Carli, G. (1993) Cinquant'anni di vita italiana. Bari, Laterza

Cassese, S. (1999) Tavola rotonda. In: Amato, G. Cassese, S. Turchetti, G. Varaldo, R. Eds Il governo della sanità.Milano, F.Angeli, pp. 129–131

Cazzola, G. (1997) La sanità liberata: Il mercato possibile per la tutela della salute. Bologna, Il Mulino

Confindustria (1997) Proposte per una nuova sanità. Roma, Mimeo

Della Sala, V. (1997) Hollowing out and hardening the state: European integration and the Italian economy. West European Politics, 20 (1), pp. 14–33

DOH—Department of Health (1997) The New NHS. Modern, dependable. London, HMSO

DOH—Department of Health (1998) A First Class Service. HMSO

Dyson, K., Featherstone, K. (1996) Italy and EMU as a "vincolo esterno": empowering the technocrats, transforming the state. South European Society & Politics, 1 (2), pp. 272–299

Enthoven, A.C. (1985) Reflections on the management of the NHS. London, Nuffield Provincial Hospital Trust

Fabbrini, S. (2000) Political change without institutional transformation: what can we learn from the Italian crisis of the 1990s? International Political Science Review, 21 (2), pp. 173–196

Falcitelli, N., Langiano, T. (2004) (a cura di) Politiche innovative nel SSN: i primi dieci anni dei DRG in Italia. Bologna, Il Mulino

Fattore, G., Jommi, C. (1998) The new pharmaceutical policy in Italy. Health Policy, 46 (1), pp. 21–41

Ferrera, M., Gualmini, E. (2004) Rescued by Europe? Italy's social and labour market reforms from Maastricht to Berlusconi. Amsterdam, Amsterdam University Press

Fetter, R.B., Shen, Y., Freeman, J.L., Averill, R., Thompson, J.D. (1980) Case-mix definition by Diagnosis-Related Groups. Medical Care (suppl.) 18 (2), pp. 1–53

Fiorentini, G. (2000) Società a capitale misto nell'offerta di servizi sanitari. Mercato, Concorrenza, Regole, 1, pp. 85–111

France, G. (1990) (a cura di) Il Libro Bianco "Al servizio dei pazienti". La riforma del Servizio Sanitario Britannico. Una valutazione Italiana. Roma, CNR—Istituto di Studi sulle Regioni. Quaderni per la Ricerca—serie studi n. 20

France, G. (1997) Sanità: le insidie dell'universalismo. Il Mulino, 369 (1), pp. 170–181

France, G. (1999) Concorrenza in sanità: L'opinione dei Direttori Generali. In Politiche sanitarie in un sistema di governo decentrato: Il caso della concorrenza nel SSN. France G., Ed. Milano, Giuffrè, pp. 213–244

France, G., Taroni, F. (2000) Starting down the road to targets in health: The case of Italy. European Journal of Public Health, 10 (4), pp. 25–29

France, G., Taroni, F. (2005) The evolution of health-policy making in Italy. Journal of Health Politics, Policy and Law, 30 (1–2), pp. 169–188

France, G., Taroni, F., Donatini A. (2005) The Italian health-care system. Health Economics, 14, pp. s187–s202

Garattini, S. (1995) Cultural shift in Italy's drug policy. Lancet, 346, 5–6

Gaynor, M., Ho, K., Town, R.J. (2015) The industrial organization of health care markets. Journal of Economic Literature, 53 (2), pp. 235–284

Goddard, M., Mannion, R. (1998) From competition to co-operation. Health Economics, 7, pp. 105–119

Gonnella, J.S., Hornbrook, M.C., Louis, D.Z. (1984) Staging of disease: a case-mix measurement. Journal American Medical Association, 251 (5), pp. 637–644

Gorsky, M. (2013) "Searching for the people in charge": Appraising the 1983 Griffiths NHS Management Inquiry. Medical History, 57 (1), pp. 87–107

Gorsky, M., Millward, G. (2018) Resource Allocation for Equity in the British National Health Service, 1948–89: An Advocacy Coalition Analysis of the RAWP. Journal of Health Politics, Policy and Law, 43(1), pp. 69–108

Gourevitch, D. (1978) The second image reversed: the international sources of domestic politics. International Organization, 32 (4), pp. 881–912.

Hall, P.A. (1993) Policy paradigms, social learning and the state: The case of economic policy-making in Britain. Comparative Politics, 25 (3), pp. 275–296

Istituto di Studi sulle Regioni (1990) "Al servizio dei pazienti". Il Libro Bianco sulla riforma del Servizio Sanitario Britannico ("Working for patients" The White Paper on reform of the British NHS), Quaderni per la ricerca—serie documentazione 8. Roma, Istituto di Studi sulle Regioni.

Kingdon, J.W. (1994) Agendas, alternatives and public policy. New York, Harper & Collins

Klein, R. (2006) The new politics of the NHS: From creation to reinvention. Oxford, Radcliff Publishing

Le Grand, J. (1991) Quasi-markets and social policy. Economic Journal, 101, pp. 1256–1267

Light, D.W. (1997) From managed competition to managed cooperation: theory and lessons from the British experience. The Milbank Quarterly, 75 (3), pp. 297–341

Marmor, T. (2004) Fads in medical care management and policy. London, TSO

Marmor, T.R., Plowden, W. (1991) Rhetoric and reality in the intellectual jet stream: the export to Britain from America of questionable ideas. Journal of Health Politics, Policy and Law, 16, pp. 807–812

Mattei, P. (2007) Legislative delegation to the executive in the "second" Italian Republic. Modern Italy, 12 (1), pp. 73–89

McCarthy, P. (1995) The crisis of the Italian state: From the origins of the cold war to the fall of Berlusconi. Basingstoke, Macmillan Press

Minerva, D. (2009) La fiera delle sanità. Milano, Rizzoli

Onofri, P. (2008) La Commissione per l'analisi delle compatibilità macroeconomiche della spesa sociale. In: Guerzoni L. (a cura di) La riforma del welfare: Dieci anni dopo la "Commissione Onofri". Bologna, Il Mulino, pp. 43–58

Osborne, D., Gaebler, T. (1992) Reinventing government: How the entrepreneurial spirit is transforming the public sector. Reading, MA, Addison-Wesley

Pierson, P. (2004) Politics in time. History,institutions, and social analysis. Princeton, Princeton University Press

Piperno, A. (1997) Mercati assicurativi e istituzioni: La previdenza sanitaria integrativa. Bologna, Il Mulino

Pollock, A.M. (2004) NHS plc: The privatization of our health care. London, Verso

Quaglia, L. (2005) Civil servants, economic ideas and economic policies: Lessons from Italy. Governance, 18 (4), pp. 545–566

Rhodes, M. (2015) Tangentopoli—More than twenty years on. In: The Oxford Handbook of Italian Politics. Jones E., Pasquino G., Eds. Oxford, Oxford University Press, pp. 309–323

Saltman, R., Von Otter, C. (1991) Saggi sulla teoria della competizione pubblica nel settore della Sanità. Roma, Formez

Scarpa, S., Chiti, L. (1975) Di farmaci si muore. Roma, Editori Riuniti

Skocpol, T., Finegold, K. (1982) State capacity and economic intervention in the Early New Deal. Political Science Quarterly, 97 (2), pp. 255–278

Smith, R. (1999) Perfidious Financial Idiocy. A "free lunch" that could destroy the NHS. British Medical Journal, 319, pp. 2–3

Taroni, F. (1996) DRG/ROD e sistemi di pagamento degli ospedali. Roma, Il Pensiero Scientifico

Taroni, F. (2001) Livelli Essenziali di Assistenza. Sogno, Miraggio o Nemesi? In: Fiorentini, G. Ed. I servizi sanitari in Italia, 2000. Bologna, Il Mulino, pp. 27–91

Timmins, N. (1996) The five giants: A biography of the welfare state. London, Fontana Press

Watts, R.L. (1989) Executive Federalism: A comparative analysis. Institute of Intergovernmental Relations, Queen's University

CHAPTER 9

New Issues at the Dawn of the Twenty-first Century

The 1990s represent an historic turning point in Italian politics (Newell 2020). The dramatic events triggered a political and institutional crisis that ended the postwar "old regime" and started a complex transition from the "First" to the so-called Second Italian Republic (Bull and Newell 2005, p. 2). Scholars disagree as to whether this difficult passage had been completed (Bull and Rhodes 1997), and some even argue that it "never happened" (Gentiloni Silveri 2015). All recognize however that by the early 1990s the fundamental characteristics that shaped postwar Italian politics were no longer present and that the ongoing "transformation of the *generality* is composed of differential degrees of change at the *specific* sectorial level" (Bull and Newell 2005, p. 3, emphasis in the original). In health policy and politics, a transition certainly occurred and started much earlier than in the other sectors. In 1974, the postwar social insurance system based on the *casse mutue* ended. In 1978, the creation of National Health Service (SSN) triggered a sweeping institutional transition. Providing universal access to uniform services, the SSN broke with two fundamental characteristics of the Italian welfare system, its particularistic nature and the provision of cash instead of services (Ascoli 1984). Moreover, beyond the rules governing benefits and eligibility, the SSN introduced new structural issues relating to financing, based on taxation instead of contributions, and governance, with the statutory participation of municipalities and the increasing role of the regions, instead of the centralized rule by the national institutes. These radical institutional changes

F. Taroni, *Health and Healthcare Policy in Italy since 1861*,
https://doi.org/10.1007/978-3-030-88731-5_9

set the healthcare state and its dynamics apart from the rest of the welfare state.

In the 1990s, the SSN underwent the biggest change in its history. At the end of the century, its structure and functioning were remarkably different from its beginning, although its fundamental principles remained notionally in place. These changes shaped the SSN for the next twenty years. The dominant problems and proffered solutions in the new policy agenda of the 1990s differed greatly from the ideas and issues that had figured prominently in the debate over the institution of the SSN. How these ideas developed and were implemented (or not) are the main objects of this chapter, which takes a broader view than the structure and organization of health services examined above. It focuses on the transition from a policy of expanding access to an ever-enlarging range of benefits to the modern politics of curbing excess in health expenditures and individual health behaviour, including a new interpretation of prevention. It also considers the evolution from community control of health and healthcare governance envisioned in the 1978 law to the adoption of a standard model of customer satisfaction adopted by *aziende sanitarie* in the 1990s. This largely set the stage of the main issues in the health policy arena of the twenty-first century. However, it is first necessary to examine the response to the abrupt epidemic of HIV/AIDS infection, the greatest challenge to the health of the nation before the age of COVID, which put to test the new SSN in the making under conditions of extreme financial duress.

The First Two Decades of the HIV/AIDS Epidemic

The first Italian case of AIDS was diagnosed in 1982, in a man having sex with men and a history of frequent travel in the United States (Taroni et al. 2007). For some years, the scale of the epidemic remained very small, and AIDS was conceived as a "gay plague" by analogy to the sudden, time-limited epidemics of the past. By the end of 1983, nine cases had been reported. The next year AIDS cases doubled, and for the first time one of such cases was endogenous, an intravenous drug user, heterosexual, with no history of travelling abroad. It soon became apparent that in Italy intravenous drug users were a major at-risk group. By the end of 1987, WHO statistics showed in Italy a cumulative total of 1104 cases, a rate of 19 cases per million people, well below France (46) and Denmark (40) and lower than Britain (22). The United States' rate at 209 cases per million people offered a threatening vision of a possible future to all European countries (Steffen 1993). Two-thirds of people with AIDS in Italy were drug addicts, a feature of the epidemic that increased the risk of spread to the general population through sexual transmission. This added a further level of uncertainty to the forecasts due to inadequate knowledge of fundamental parameters such as the distribution of sexual preferences, the frequency of different types of intercourse, the number of partners, all elements hitherto far from the "normal" attention of clinicians, epidemiologists, and health planners. And in fact, by the end of 1990, cases increased to 8227, and the rate climbed from 25 to 143 cases per million population, second only to France among European countries.

The universal and comprehensive principles of the recently established SSN notionally secured access to preventative services and acute and long-term care for everyone infected with HIV/AIDS. The principles filled the cracks of the previous social insurance systems based on *casse mutue*. The occupationally based system would have left many gaps with a young, frequently unemployed population, in terms of who was covered, for how long, and for what services. Moreover, the high cost of care would have exacerbated the shaky fiscal basis of *casse mutue*. All these coverage and cost issues gained urgency as effective and very expensive drugs became available, such as AZT in 1987 and Highly Active Anti-retroviral Therapies (HAARTs) in 1996. The recent introduction of the SSN and the uncertain knowledge about the disease and unpreparedness of the hitherto neglected area of infectious disease explain the relatively slow response of the health system, in Italy as in other countries (Ferlie 1993; Steffen 2004). In the early phase of the epidemic, the quality of clinical care was poor (Taroni and Anemona 1993). Patients treated in the principal Italian centres experienced recurrent hospital admissions with life-threatening conditions. The first diagnosis of the disease was often at advanced stages of the disease, and programmes for preventing opportunistic infections were limited. The length of hospital stay was extended at all stages of the disease; there was a lack of alternatives to hospitalization, including ambulatory and day stay for acute cases and hospice for palliative and terminal care.

AIDS became a national political priority in 1986–1987, the turning period for public policies in several European countries, including Britain (Day and Klein 1989) and France (Steffen 1993). The fight against HIV infection was given special status, and policies were managed by *Commissione Nazionale AIDS* (CNA—National AIDS Committee), a panel of clinicians, researchers, and public health experts, which included patient representatives, extraordinary for Italy. The Minister of Health chaired the CNA, but its activity was under the technical leadership of Elio Guzzanti, a long-time vice-president of the Commission, and then Minister himself. Guzzanti boasted long experience and a deep knowledge of both public health and health planning. Under his leadership, the government developed a national programme against HIV infection that was essentially a "policy of experts" as in most European countries (Fox et al. 1989).

The CNA, along with the AIDS Operations Center at the *Istituto Superiore di Sanità* and a collaborative network of centres throughout the country, managed initiatives in four main areas: prevention, service organization and planning, health education, and research. Prevention was initially the only tool to fight HIV infection and save lives. This required reinforced interventions in the field of public health, which was still at the margin of the medical sector and required funding and institutional support. Prevention programmes however involved politicized choices about sensitive personal issues exposed to political and moral attacks, particularly in a Catholic country. Sex education, needle exchanges, and HIV compulsory testing and reporting were particularly controversial. Moral arguments about sexuality and drug use were an important feature of the HIV prevention debate. The use of condoms in homosexual and heterosexual extramarital intercourse and needle exchange programmes for intravenous drug users were represented as conveying incentives to engage in immoral, and even illegal, behaviour. Ideological discourses about sexuality and drug use were a constant feature of the Italian debate. These found important supporters in the more conservative part of the Church, while several of its organizations participated instead in preventative programmes and terminal care for drug users. In 1988, every home in Italy received a letter signed by the Health Minister Carlo Donat Cattin supporting abstinence as the policy for preventing HIV/AIDS. Public health experts and organized advocacy groups actively campaigned for harm reduction programmes, which were frequently funded locally from *comuni* and *provincie* as part of their social services. Harm reduction strategies were framed, in the main, more as interventions to protect the health of the general public than to limit the damage to drug users, as with the McClelland Committee report of 1986 in Edinburgh (HIV infection in Scotland 1986). With the main message "if you know it, you avoid it", a bold and massive campaign aimed at the general population through billboards, posters, and TV spots with the goal of raising general awareness about the AIDS problem. The health propaganda campaign was unprecedented in scale and in its unusual frankness, even if did not explicitly addressed sexual conduct.

Law nr. 35 of 1990 was the instrument CAN used to adapt the structure and functioning of the SSN to the emerging needs of treating the increasing number of patients with AIDS. The programme addressed both the well-established hospital sector and the weakly structured medical and social services for drug users. Hospital interventions included expanding the capacity of the infectious disease sector, which had been widely curtailed in the assumption that contagious diseases were no longer a problem of modern healthcare systems. Restructuring available resources and building new structures were planned to meet an estimated capacity of over 20,000 hospital beds needed against the slightly less than 7000, which were currently available. They would include new day-hospitals for outpatient treatments of acute cases as well as hospice facilities for palliative and respite care of terminal patients. The programme also helped bring interventions for drug users into the mainstream and integrated them into the "normal" range of medical services provided by USLs for the first time. The law also imposed a strict prohibition of compulsory screening for HIV infection, particularly in the workplace, which indirectly stressed emphasis on voluntary testing, counselling, and reporting.

Finally, CNA launched a massive research investment through a strategy based on the model of national competitive projects. The National Research Program on HIV infection coordinated by the *Istituto Superiore di Sanità* made HIV/AIDS a major area of publicly funded research that set research priorities in different sectors, including basic and translational medical research, as well as social, behavioural, and economic research. This greatly accelerated the diffusion of clinical competence and quality of care and contributed to the spread of best professional practices and developing guidelines for the treatment and prevention of infection, including in the workplace and among health professionals.

Annual deaths from AIDS continued to rise steadily, with new cases peaking at 4500 and 5033 in 1995 and 1996 respectively. Breakthroughs in medical treatment shifted the focus from palliative care to effective treatment in halting the disease progression through the advent of protease inhibitors and HAART. The SSN ensured people with HIV infection timely access to services that the advancement of knowledge made gradually available. From that point on, AIDS began a sharp decline to 527 deaths and 2090 new cases in 1998 and continued thereafter, although the size of the HIV-positive population continued to grow (Ministero della Sanità 1999). The availability of effective long-term treatment transformed AIDS into a manageable chronic condition and incentivized routine testing and early diagnosis (Ippolito et al. 2001). HIV infection was also progressively distanced from a condition exclusively of socially stigmatized groups, such as gay men and inject drug users, marshalling the "innocent victim" discourse of patients infected by blood transfusions. Changes in medical practice and social perception shifted towards treating AIDS as "just another medical condition". HIV infection came to be seen less a sudden and time-limited epidemic than an endemic, long-standing condition (Fee and Krieger 1993). Forecasting also adapted to the new disease model. Assumptions of exponential growth were dropped, and the AIDS problem was seen less in dramatic and immediate terms than a continuing threat into the twenty-first century. The national AIDS programme remained in place, but its implementation increasingly focused on the role of the regions that had become the main actors of general health policy.

The HIV/AIDS epidemics posed novel problems characterized by scientific uncertainty and moral ambiguity, as it affected mostly marginalized and stigmatized groups. It raised significant debate in many areas of Italian society, including the professional community and the Catholic Church. The political response to the epidemic in Italy reveals more similarities than differences with other countries, including France, Britain, and the United States (Fox et al. 1989; Berridge and Strong 1993; Steffen 2004). An important difference from United States, where access and financial problems were paramount (Padamsee 2018), and a strong similarity with Britain, was the existence in Italy of a universal and comprehensive health system, although still in the making and in a period of financial duress. The year 1982 when the first Italian case of AIDS was diagnosed was also the year of passage of a budget that so severely cut the national health fund to raise doubt over the implementation of Law 833. Financial problems increased in the second half of the 1990s with the advent of HAART Therapy. Prevention problems remained politicized, particularly with regard to needle exchange and other harm reduction programmes for drug abusers. As in most other countries, HIV/AIDS policy in Italy was both "a series of responses to the concrete challenges of a health crisis" as well as "a resource used to wage broad social and ideological battles" (Padamsee 2018, p. 1001). Governments initially seemed to ignore the problem, and then created a national programme which was specifically focused on fighting HIV infection and was essentially based on "the authority of conventional medical and public health leaders" (Fox et al. 1989, p. 107). Centralizing the development of AIDS policy in an expert committee at the national level helped create an elite consensus over technical matters. This insulated HIV/AIDS policymaking from political and ideological pressures as well as from the turmoil that general health policy went through during the 1980s and the 1990s.

From the Politics of Access to the Policy of Excess

The SSN never enjoyed the period of consolidation that Klein acknowledges for the British NHS in the first ten years of its life (Klein 2006, p. 49). Instead, at the very time of its institution the SSN had to adapt to a regime of fiscal duress and to face the rising expectations and demands of a consumer society that tended to feel that its services, if not its principles, were inadequate for the time. Further developments brought about significant differences with the ideas that had been debated at its institution. For many of its supporters, the SSN had come to epitomize if not "a piece of real socialism" as Aneurin Bevan had called the British NHS years before (Bevan 1950), at least an institution incorporating the ideals of social solidarity and distributional justice, ensuring universal access to health services based only on need. In this sense, the institution of the SSN was the outcome of a redistributive politics aiming to expand access to publicly funded services. The ideas behind the politics of access had also developed some specifics about the nature of the services to be offered. The ideas of deinstitutionalization, for example, prioritized care in the community and by the community, and focused on primary social and healthcare by professionals as well as from lay providers. Among the topics figured prominently the concept of prevention, which was one of the ideological planks of the trade unions. Primary prevention was to be a collective commitment oriented towards removing the root causes of diseases both in the workplace and in the community, and this was prioritized over secondary prevention, or early detection of diseases, dubbed at times *falsa prevenzione* (false prevention) (Maccacaro 1976). Corporatization, managerialism, and cost control were instead the dominant themes of the debate in the 1980s and the 1990s. Prevention was narrowly redefined by the New Public Health to mean an individual's responsibility for his or her own health.

Julien Le Grand provides useful insights to help understand the evolving concepts and the intertwining between the policy of health and of healthcare services under the common logic of curbing excess consumption (Le Grand 2008). Its arguments revolve around the contrast between

Beveridge's aims at eliminating the problems of "too little" and the modern health problems which arise instead from excessive consumption of various kinds: "people drink, smoke and eat too much. They take illegal drugs; they eat the wrong kind of food, and they undertake too many sedentary activities" (p. 854). While cholera, typhoid, and diphtheria arose from want and misery, diabetes, obesity, stroke, and the like are problems of "too much" and originate from individual choices and behaviours. Health promotion should therefore adopt the philosophy of libertarian paternalism (Sunstein and Thaler 2003) to curb unhealthy activities without interfering with individual freedom and autonomy. Health promotion would "nudge" people into appropriate behaviours to promote their long-term well-being and save cost to the healthcare system (Thaler and Sunstein 2008).

The politics of excess and the logic of addressing individual's behaviours to remedy systemic problems also apply to health services. From the early 1980s, health services research devoted increased attention to the problem of "too much" medicine, whereby patients "consume" too many tests and treatments providing them little or no benefits. Variation studies have shown substantial "unwarranted" differences in the use of various healthcare services that cannot be explained by patient illness or preferences (Wennberg and Gittelsohn 1973). In the United States, federal agencies such as the Office of Technology Assessment (OTA), the Health Care Financing Administration (HCFA), and the Agency for Health Care Policy and Research (AHCPR) (now with different names: see glossary) participated in developing a host of regulatory instruments, in collaboration with medical research organizations. Researchers at Rand Corporation pioneered a series of "appropriateness studies" to examine the reasons for high- and low-use areas, and answer Wennberg's famous question "are hospital services rationed in New Haven or over-utilized in Boston" as two areas of respectively low and high rates of hospital admissions (Wennberg et al. 1987). The Rand approach classified health services into four categories of appropriateness, ranging from "necessary" (whose lack implied "underuse") to "inappropriate", which implied overuse (Brook 1995). This was extended to examine the appropriate and/or timely use of specific services, such as both hospital admissions and the appropriateness of each day of a hospital stay, as with the Appropriateness Evaluation Protocol (AEP) (Gertman and Restuccia 1981).

In the United States, such instruments were used to regulate supply, use, and prices of health services by federal and state programmes as well as by private insurance companies. Certificate of Needs (CON) authorized the building of new hospitals or the expansion of the older ones, and the acquisition of big technologies, including, for example, CT-Scans (Dranove 2000, p. 53). Utilization Review (UR) was extended by private health insurance companies from its original federal application to the retrospective review of Medicare patients' hospitalizations (the so-called Professional Standard Review Organizations—PSROs) to authorize common, high-volume tests and procedures (Coast 1996). Control of unit prices was initiated with Medicare's adoption of the hospital Prospective Payment System (PPS) based on DRGs (Mayes 2007) in 1983 and later extended to physicians' services through the Resource-based Relative Value system of payment (Hsiao et al. 1988). Scientific research, professional organizations, and payment policies aligned to create methods, tools, and incentives to limit overuse of unnecessary and of inappropriate medical services in what has been defined a form of "technocratic corporatism" (Brown 1985). Doctors faced mounting pressures to use procedures only when science-based criteria and clinical guidelines developed by federal agencies, private research organizations, and/or professional organizations indicated they were clinically appropriate.

The widespread adoption of schemes of micro-regulation of individual physician's practices contrasts with the state of Oregon's scheme of rationing medical benefits for its population assisted with Medicaid (Oberlander et al. 2001). To allocate the limited resources available, the Oregon Health Services Commission developed a priority list of funded treatments, which ranked 714 condition-treatment pairs based on a cost-utility formula incorporating also public input. Oregon's "bold idea" of "rational rationing" medical resources was more in line with European experiences of macroeconomic control of health usage and expenditures. The idea of "drawing the line on Medicare coverage", that is, removing items from the list of covered services, particularly attracted interest from health systems such as Canadian Medicare, the British NHS, and the Italian SSN, which were debating priority setting for public coverage of medical services under strict financial constraints. Paul Pierson's seminal concepts of the "new politics" and of the "retrenchment" of the welfare state (Pierson 1994, 1996) first focused on the differences with the "old" expansive policy in terms of the means adopted and the configuration, which resulted. Cutbacks had to consider both welfare states' support from interest groups ready to mobilize and their own institutional inertia, which allowed only incremental change over time. Both factors explain why politicians and reformers abstained in the main from radical and highly visible cutbacks across the board and adopted instead selective policies addressing specific services and/or users' categories. Moreover, to deflect blame, reformers engaged in policies of compensation, obfuscation, and divisions (Weaver 1986). The implementation of the new politics of taking away "excessive" benefits accounts for the insular configuration of the retreat, that is, "the retrenchment" of the welfare state, which focused in selected sectors, services, and/or groups of beneficiaries. This exposed the new phase of health policies no longer aimed at expanding entitlement to health services (the politics of expanding access) but to contain their consumption and curb health expenditures (the politics of containing excesses). Services excluded from coverage were frequently taken up by allegedly "integrative" or "complementary" programmes piggybacking statutory provision. This gave rise to the current process of "hybridization" in the structure of the health systems, which incorporate elements taken from different models departing from their original ideal types (Tuohy 2012).

Italian reformers have reacted differently in addressing the diseases of excess and in implementing organizational changes to curb "excessive" service consumption. Typical "giants of excess" such as obesity and diabetes, frequently described in "epidemic" terms, have generally gained relatively low political salience. Both conditions are still perceived more as personal medical problems best suited for individual clinical treatment than collective problems deserving specific interventions. Several Italian governments have not been shy about intervening in ostensibly private behaviours, such as the abuse of alcohol, drugs, and tobacco as well as family planning and fertility control. However, one can find very few of the seven "triggers" that usually raise the social and political salience of personal health issues and prompt government intervention in private habits (Kersh and Morone 2002). Social and self-help movements that in the 1970s mobilized for abortion clinics and the pill are virtually non-existent in Italy against, for example, obesity, where the focus is mostly on surgical treatment. Medical warnings against the dangers of tobacco and the abuse of alcohol and drugs have however stimulated health promotion and prevention programmes and greatly contributed to legal bans on smoking in enclosed spaces, advertising and sponsorship, and so on. An important exception is the "unexpected success" of the smoking ban which was implemented in 2005 (Mele and Compagni 2010). Reasons could be found in the doctors' lukewarm reception of the New Public Health emphasis on self-control of personal lifestyles and their difficult transition from traditional medical paternalism to patients' autonomy and personal empowerment. Lacking the mobilization of public opinion and of widespread support from the health professionals, feeble and sparse government attempts to regulate the so-called Fat Industry have so far easily resisted, and proposals for taxing high-fat foods and high-sugar drinks quickly dropped. Of course, the Ministry of Health has produced a number of policy papers recognizing the association between food, nutrition, and health; warning against low-nutrition, high-fat foods; and promoting the virtue of the Mediterranean diet. However, to date governmental action on food consumption remains essentially anchored in the main to the Old-Public Health and focused on the traditional issues of purity and safety. 2010).

In regulating healthcare organization, reformers in Italy had adopted both traditional, across-the-board measures and new selective policies. In the 1980s, a systematic budgetary policy of underestimating expenditures and overestimating savings (in itself, an exercise of blame shifting from central government to the regions) maintained the SSN in a state of chronic instability and financial crisis. In the early 1990s, healthcare was addressed as one of the sectors contributing to excessive public expenditures, along with pensions, local finance, and public administration. Corporatization and managerialism were introduced as means for restoring accountability, increasing efficiency, and cutting waste, the traditional rhetoric tropes for policies of obfuscation. Devolving to the regions the responsibility for overrunning their allocated budgets was again a classical example of blame shifting. A key indicator of the success of the strategy was in reducing the levels and containing the dynamic of public health expenditure. In 1990, public health expenses' share of the gross domestic product was 6.3 per cent; in 1995, it fell to 5.4 per cent and remained stable with an increasing GDP until 1999, when it rose to 5.7 per cent of GDP (Ministero della Sanità 1999). This was the combined effect of cutting back entitlements at the central level in a policy of "conditional universalism" (Ferrera 1995) and streamlining the delivery of health services at the local level through a policy of clinical and organizational appropriateness and monitoring professional behaviours. A technocratic project of "governing through evidence" combined central planning and local accreditation with incentive provision and the activation of professional networks (Mele et al. 2013).

A surge of transnational learning introduced in Italy methods and tools for utilization review and quality assurance, as well as the new concepts of evidence-based medicine and clinical guidelines. A relatively limited capacity for health service and evaluation research was built at the national and regional level of the SSN (such as, for example, the *Istituto Superiore di Sanità*), as well as in non-profit private organizations such as the Istituto Mario Negri in Milan and Bergamo (Light and Maturo 2015). The Institute embeds patient and public health oriented clinical and epidemiologic research exploiting the SSN as a "laboratory" (Rovelli and Tognoni 1996) aiming at changing attitudes and practices of participants in mega-trials, such as GISSI (Tognoni et al. 2019). Legislation instituted a national programme of compulsory accreditation of all public hospitals and health organizations, including private facilities willing to provide services paid by the SSN. A culture of quality assessment and improvement was promoted within the ASLs and developed extensively in certain clinical areas such as oncology, obstetrics, and cardiology, and particularly among nurses. More information became available to managers, patients, and the public about outcomes of procedures and doctors' behaviour through performance indicators, resource management, and medical audit. Through their national organizations, health professionals in various specialties took greater responsibility for regulating their standards of practice (see for example Schweiger and Scherillo 2000 for cardiology). This helped maintain professional control of the process and prevented the regulations from becoming only managerial tools. Monetary incentives were deployed to increase participation in local projects, and doctors and other health professional, particularly nurses, faced mounting pressures for the appropriate use of resources. In a significant difference from American experiences, regulations and incentives were designed to improve physicians' clinical competence and not used to curtail patients' benefits or regulate access to services. However, this was not the case with central provisions regulating patient co-payments over drug prescriptions, physicians' specialist visits, tests, and treatments.

Cost-sharing policies concerning the use of medical services are a traditional feature of the SSN. Limiting moral hazard, that is, making service users economically responsible and preventing free riding over SSN common pool resources, is the altruistic aim officially claimed. Less frequently voiced, but of great fiscal importance, is their contribution to increase the revenues to the SSN coffers. Out-of-pocket expenditures imposed on services' users, however, also have a significant, if indirect, impact over universal and fair access to services. This is how cost-sharing policies can be instrumental in producing a form of "conditional" (Ferrera 1995) or "selective" (Onofri 2008) universalism. The WHO "cube" model describes universal health coverage along the three axis of population coverage (its breadth), service coverage (its depth), and financial coverage, or the protection of the population against economic barriers to access (WHO 2010). This explains how "selective" or "conditional" coverage entails available benefits, eligible populations, and the distribution of public and personal, private cost. Excluding from coverage a segment of the population is a rare event in the history of the SSN. It was legislated just once, in September 1992 and then immediately rescinded within a month for a softer version where permissive legislation allowed the regions to let part of their citizens exit SSN coverage in case of overrunning their allocated budgets. As we have seen, this version was also ended in a few months (see page 192). Cost sharing is the typical means of the potential path to "selective" universalism in the SSN. High out-of-pocket expenses to access an SSN service can either block access to benefits notionally available but economically unsustainable for the less well-off, or incentivize the well-to-do to exit the SSN for the private market when cost and convenience make it preferable. Therefore, it is important to consider the three dimensions of co-payment policies comprising the amount of co-payment, the services included, and the ceiling and exemptions allowed.

Measures adopted in the early 1980s increased the amounts of co-payments, expanded the services included, and reduced exemptions. All drugs and specialist services required co-payments, including outpatient specialist visits (but not those from general practitioners), tests, and treatments. Hospitalization had always been excluded; by the mid-1980s, a decree imposing a boarding charge to hospital patients provoked an uproar and was quickly retracted. Exemptions were initially based on age and income, and then also on services for some chronic conditions, such as hypertension, diabetes, and COPD. For drugs, the balance between increased cost sharing, the number of exempt people (which went from 18 to 25 per cent of the population), and their disproportionate drug use (which increased from 45 to 75 per cent of the total) maintained their revenue at about 10 per cent of total pharmaceutical expenditures (Ferrera 1995, p. 290). Private health spending increased from 18.2 per cent to 20.2 per cent of total health expenditure, or about 1.22 per cent of GDP. The expected exit from the SSN and "defection" to the private market was not observed in the few studies performed (Piperno 1986). The 1990s' comprehensive and wide-ranging revision of the national formulary satisfied the political mandate to squeeze public spending on drugs below the previously established ceiling and increased private expenditure, as physicians maintained their prescription behaviour (see page 197).

Regions responded differently to the great transformations in the healthcare sector of the 1990s, which affected the equity goals of the SSN. Autonomy favoured the more dynamic northern regions, and this further increased the already significant gap with the southern regions in all the relevant performance indicators. Health expenditures per capita were higher in the northern regions, but budget deficits concentrated in the southern regions. Satisfaction with healthcare, although generally low (see below), was higher in the northern than in the southern regions. Approximately half a million patients per year moved from the southern regions to the hospitals of the northern regions. Lombardy and Emilia-Romagna were particularly attractive, while the regions with the worst net balance between incoming and outgoing hospital patients were Campania, Calabria, and Sicily. As in interregional mobility the money followed the patients, incoming mobility was a source for additional income for the northern regions, while those with a negative balance of cross-border mobility paid both for their internal inefficiencies and for the cost of their citizens' exit (Brenna and Spandonaro 2015).

Responding to Public Expectations: Citizens' Views

Public dissatisfaction with the health system in Italy was quite high before the establishment of the SSN and continued to be high in all surveys that have been conducted since. In a survey of public evaluation of governmental performance in medical care conducted in several countries between 1973 and 1976 (no precise date is given for the Italian poll), Italy showed the most negative evaluations. About two-thirds of the sample of 1701 people interviewed rated government's performance either "bad" (39.0 per cent) or "very bad" (22.6) compared, for example, to 14.7 per cent in Britain (Pescosolido et al. 1985). Higher education groups and those who considered "medical care an important issue" gave worse performance ratings. In a survey performed in the early 1990s, before the passing of the first healthcare reform, only 12 per cent of respondents, the lowest percentage in the countries surveyed, believed that only minor changes were needed to make healthcare work better (Blendon et al. 1990). About 86 per cent of Italian responders, as well as 69 per cent of Britons, were in favour of a major overhaul of the health system. In the Eurobarometer survey performed in 1996, after the first major health reform, 76.9 per cent of responders in Italy were still asking for major reform, compared to 56.0 per cent in Britain (Mossialos 1997). The question about satisfaction is rather vague and may reflect a general distrust of government and its institutions, while asking citizens about their support for fundamental changes in the medical system seems more specific and to the point. A further survey performed at the end of 1998 on a sample of over 10,000 citizens confirmed the overall dissatisfaction with the system in over 60 per cent of the responders but provided also some additional insights (Ministero della Sanità 1999). Satisfaction was much higher among those who had used the system in the last six months, among the elderly and some vulnerable groups, as well as for specific services, particularly related to primary care. Such differences in the citizens' views between a negative image of the health system in general and the positive experiences of those who had used its services seem to suggest that dissatisfaction reflects the lack of public confidence in institutions in general rather than problems with medical care. It also suggests that over the years the SSN, or at least some of its programmes, had built up a constituency of their own, which was in favour of the status quo and could possibly mobilize to resist change.

General surveys and opinion polls are of some use in helping to understand the climate under which systems operate and assessing the support that existing programmes have from various population groups. More is needed to seek democratic participation, as the original 1978 legislation claimed, empower patients or, at least, listen to their voices. Italy did not implement the vague aspirations of public participation in local health policymaking. It did not either issue a national Patients' Charter or engage with the policy of Patient choice that became a standard objective of the British NHS (DoH 2003). *Aziende sanitarie* were legally asked to develop a complex apparatus of programmes of customer satisfaction, including performing periodical surveys of patients' opinions and "perceived quality" of services, managing their complaints, and, particularly, issuing their own *Carta dei Servizi* (Health Services Chart). *Carta dei Servizi* was an official and legally required document of private and public health organizations, documenting performance goals and indicators, including waiting times, quality assurance programmes, and the process for patient complaints and their management (Cinotti and Cipolla 2003). In essence, *Carte dei servizi* provided service users the opportunity to complain about the deviation from a confused mix of customer service standards specific to each *azienda sanitaria*. There are a number of reasons why the SSN settled for routinely adopting standard methods and tools of customer relations instead of pursuing aggressive strategies of patient empowerment. First, most doctors continue to be seen, following Le Grand's famous metaphor, as kings who take their patients' well-being to heart and not as knaves, or income- and rent-seeking bureaucrats (Le Grand 2003). Another possible reason is that Italy never saw the rise of a strong consumerist movement, particularly in the health field. It saw instead the mushrooming of single-issue groups, practising "healthism" or "health for me" (Greenhalgh and Wessely 2004), close to what Muir Gray dubbed post-modern medicine: "a distrust of science, a readiness to resort to litigation, a greater attention to risk and better access to information of whatever quality, not necessarily associated with health literacy" (Muir Gray 1999). These were, for example, the essential traits of the highly emotional campaign promoting the Di Bella unconventional cancer therapy, which involved, in an unprecedented turmoil, the media, political parties, and all the main Italian healthcare institutions, the Parliament, and the Minister herself, just at the dawn of the new century (Remuzzi and Schieppati 1999).

References

Ascoli, U. (1984) Il sistema italiano di welfare. In: Ascoli U. (a cura di) Welfare state all'italiana. Bari, Laterza, pp. 5–51

Berridge, V., Strong, P., Eds. (1993) AIDS and contemporary history. Cambridge, Cambridge University Press

Bevan, A. (1950) In place of fear. London, William Heinemann

Blendon, R.J., Leotman, R., Morrison, I., Donelan, K. (1990) Satisfaction with health systems in ten nations. Health Affairs. 9 185–192

Brenna, E. Spandonaro, F. (2015) Regional incentives and patient cross-border mobility. Evidence from the Italian experience. International Journal of Health Policy Management, 4, pp. 363–372

Brook, R.H. (1995) The RAND/UCLA Appropriateness method. In: Mccormick A., Moore S.R., Siegel R.A, Eds. Clinical practice guidelines development: Methodology Perspectives. Rockville, MD: Public Health Service, US Department of Health and Human Services

Brown, L.D. (1985) Technocratic corporatism and administrative reform in Medicare. Journal of Health Politics, Policy and Law, 10 (3), pp. 579–599

Bull, M., Rhodes, E. (1997) Crisis and transition in Italian politics. London, Routledge

Bull, M.J., Newell, J.L. (2005) Italian politics: Adjustment under duress. Cambridge, Polity Press

Cinotti, R., Cipolla, C. (2003) La qualità condivisa fra servizi sanitari e cittadini. Metodi e strumenti. Milano, F. Angeli

Coast, J. (1996) Appropriateness versus efficiency: the economics of utilization review. Health Policy, 36, pp. 69–81

Day, P. Klein, R.(1989) Interpreting the unexpected. The case of AIDS policy-making in Britain. Journal of Public Policy, 9, pp. 337–353

Department of Health (2003) Fair for All, personal to you. A consultation on choice, responsiveness and equity. London, DoH

Dranove, D. (2000) The economic evolution of American health care: From Marcus Welby to Managed Care. Princeton, Princeton University Press

Fee, E., Krieger, N. (1993) Understanding AIDS: historical interpretations and the limits of biomedical individualism. American Journal of Public Health, 83 (10), pp. 1477–1486

Ferlie, E. (1993) The NHS responds to HIV/AIDS. In: Berridge V., Strong P., Eds. AIDS and contemporary history. Cambridge, Cambridge University Press, pp. 203–223

Ferrera, M. (1995) The rise and fall of democratic universalism: Health care reform in Italy, 1978–1994. Journal of Health Politics, Policy and Law, 20 (2), pp. 275–302

Fox, D.M., Day, P., Klein, R. (1989) The power of professionalism: policies for AIDS in Britain, Sweden and the United States. Daedalus, 118 (2), pp. 93–112

Gentiloni Silveri, U. (2015) Italy 1990–2014: the transition that never happened. Journal of Modern Italian Studies, 20 (2), pp. 171–175

Gertman, P.M., Restuccia, J.D. (1981) The Appropriateness Evaluation Protocol. A technique for assessing unnecessary days of hospital care. Medical Care, 19 (8), pp. 855–871

Greenhalgh, T., Wessely, S. (2004) "Health for me": A sociocultural analysis of healthism in the middle classes. British Medical Bulletin, 69 (1), pp. 197–213

HIV Infection in Scotland (1986) Report of the Scottish Committee on HIV Infection and Intravenous Drug Misuse (the McClelland Report). Scottish Home and Health Department

Hsiao, W.C., Braun, P., Dunn, D., Becker, E.R., De Nicola, M., Ketcham, T.R. (1988) Results and policy implications of the resource-based relative-value study. NEJM, 319, pp. 881–888

Ippolito, G., Galati, V., Serraino, D., Girardi, E. (2001) The changing picture of the HIV/AIDS epidemic. Annals of the New York Academy of Sciences, 946, pp. 1–12

Kersh, R., Morone, J. (2002) The politics of obesity: seven steps to government action. Health Affairs, 21, pp. 142–153

Klein, R. (2006) The new politics of the NHS. From creation to reinvention. 5th ed. Oxford, Radcliffe Publishing

Le Grand, J. (2003) Motivation, Agency and Public policy. Oxford, Oxford University Press

Le Grand, J. (2008) The giants of excess. A challenge to the nation's health. Journal Royal Statistical Society A, 171 (4), pp. 843–856

Light, D.W. Maturo, A.F. (2015) Good Pharma. The public health model of the Mario Negri Institute. New York, Palgrave Macmillan

Maccacaro, G. (1976) Vera e Falsa prevenzione. Sapere, 794, pp. 2–4

Mayes, R. (2007) The origins, development and passage of Medicare's revolutionary prospective payment system. Journal of the History of Medicine and Allied Sciences 62 (1) pp. 21–55

Mele, V., Compagni, A. (2010) Explaining the unexpected success of the smoking ban in Italy: political strategy and transition to practice, 2000–2005. Public Administration 88(3), pp. 819–835

Mele, V. Compagni,A. Cavazza, M. (2013) Governing through evidence. A study of technological innovation in health care. Journal of Public Administration Research and Theory, 24, pp. 843–877

Ministero Sanità (1999) Relazione sulla situazione sanitaria del paese. Roma, Tip. Stato

Mossialos, E. (1997) Citizens' views on healthcare systems. Health Economics, 6 (2), pp. 109–116

Muir Gray, J.A. (1999) Postmodern medicine. Lancet, 354, p. 1553

Newell, J.L. (2020) Italy's contemporary politics. London, Routledge

Oberlander, J., Marmor, T., Jacobs, L. (2001) Rationing medical care: Rhetoric and reality in the Oregon Plan. Canadian Medical Journal, 164 (11), pp. 1583–1587

Onofri, P. (2008) La Commissione per l'analisi delle compatibilità macroeconomiche della spesa sociale. In: Guerzoni L. (a cura di) La riforma del welfare. Dieci anni dopo la "Commissione Onofri". Bologna, Il Mulino, pp. 43–58

Padamsee, T.J. (2018) Fighting an epidemic in political context. Thirty-five years of HIV/AIDS policy making in the United States. Social History of Medicine, 33, pp. 1001–1028

Pescosolido, B.A., Boyer, C.A., Tsui, W.Y. (1985) Medical care in the welfare state. A cross-national study of public evaluations. Journal of Health and Social Behavior, 26 (4), pp. 276–297

Pierson, P. (1994) Dismantling the welfare State? Reagan, Thatcher and the politics of retrenchment. Cambridge, Cambridge University Press

Pierson, P. (1996) The new politics of the welfare state. World Politics, 48, 143–179

Piperno, A. (a cura di) (1986) La politica sanitaria in Italia. Milano, F. Angeli

Remuzzi, G., Schieppati, A. (1999) Lessons from the Di Bella affair. Lancet, 353, pp. 1289–1290

Rovelli, F. Tognoni, G. (1996) The health service as a laboratory. Lancet, 348, pp. 169–170

Steffen, M. (1993) AIDS policies in France. In: Berridge V., Strong P., Eds. AIDS and contemporary history. Cambridge, Cambridge University Press, pp. 240–264

Steffen, M. (2004) AIDS and health policy responses in European welfare states. Journal of European Social Policy, 14 (2), pp. 165–181

Sunstein, C.R., Thaler, R.H. (2003) Libertarian paternalism is not an oxymoron. University of Chicago Law Review, 70, pp. 1159–1162

Taroni, F. Anemona A. (1993) L'assistenza ospedaliera ai pazienti con AIDS in Italia. Giornale Italiano AIDS, 4, pp. 2–15

Taroni, F., France, G., Tramarin, A., et al. (2007) Italy. In: Beck E.J., Mays N., Whiteside A.W., Zuniga J.M., Eds. The HIV pandemic. Local and global implications. Oxford, Oxford University Press, pp. 476–487

Thaler, R.H., Sunstein, C.R. (2008) Nudge. Improving decisions about health, wealth and happiness. New Haven, Yale University Press

Tognoni,G. Franzosi,M.G. Garattini, S. (2019) Embedding patient- and public health- oriented research in a National Health Service: the GISSI experience. Journal of the Royal Society of Medicine, 112,pp. 200–204

Tuohy, C.H. (2012) Reform and the politics of hybridization in mature health care states. Journal of Health Politics, Policy and Law, 37(4), pp. 611–632

Weaver, K.R. (1986) The politics of blame avoidance. Journal of Public Policy, 6 (4), pp. 371–398

Wennberg, J.E., Freeman, J.L., Culp, W.J. (1987) Are hospital services rationed in New Haven or over-utilized in Boston? Lancet I, pp. 1185–1189

Wennberg, J.E., Gittelsohn, A. (1973) A small area variation in health care delivery. Science, 182, pp. 1102–1108

WHO (2010) World health report 2010: Health system financing: the path to universal coverage. Geneva

CHAPTER 10

A Lost Decade

The period following Italy's entry into the European Monetary Union was marked by the stagnation of the Italian economy and the relaxation of reform efforts in welfare policies. After the initial recovery from the crises of the early 1990s, the economy barely grew from the mid-1990s through 2008, and Italy gradually slipped behind other European countries with the second highest ratio of public debt to GDP, after Greece. A leading article in the *Economist* of May 21, 2005, dubbed Italy "the real sick man of Europe" as "Italy's economy is stagnant, its business depressed—and its reform moribund". And yet in 2001 a major Constitutional reform, confirmed by popular vote, had created the "unitary regional State" envisaged by the Constitution of 1948. While not using the term "federalism", the amendments defined the constitutional basis for a devolved system of government, where health took centre stage. A long-term plan for "fiscal federalism" and the implementation of the *Livelli Essenziali di Assistenza*, the benefit package to be uniformly provided by the SSN, consolidated the new "regional healthcare state". However, much remained to be done in terms of effective intergovernmental relations and credible fiscal rules. Stark interregional disparities and the deep divide between northern and southern regions hampered the implementation of a feasible fiscal plan, while the devolution of powers over the organization of regional services became a major political issue. This changed the way in which health policy was developed, where it took place, and its main content. Under the towering public debt, the cost containment imperative had taken a

F. Taroni, *Health and Healthcare Policy in Italy since 1861*,
https://doi.org/10.1007/978-3-030-88731-5_10

structural role. The fiscal imbalance between the central state and the regions gave the former the power of the purse. This permanently split the venues for health policymaking between fiscal policy negotiated at the central level and health service development that the constitutional reform had made a regional preserve. A new dynamic set in, until the Global Financial Crisis suddenly struck and started an entirely new process.

Regionalization, Alleged and Real

The SSN emerging from the turbulent 1990s entered the new century without the (quasi) markets that the political rhetoric had promised and without the return to the hierarchy that many had feared. The new century opened with a constitutional reform with the aim of transforming the Republic into a "unitary regional state" and placed Italy among the growing number of countries that were no longer unitary but not yet federal (Roux 2008). The amendment accorded constitutional status to the very considerable autonomy that the Italian regions had accumulated for healthcare over the preceding twenty years. Related legislation, setting the appropriate level of service coverage that should be uniformly provided by the SSN (*Livelli essenziali di assistenza*- LEA), also served the process of devolution of powers. Health had been the leading political domain where policy tools were developed and later exported to other political sectors. However, guaranteeing the fiscal autonomy necessary to very different sub-central governments in federal systems to finance functions like healthcare, which are very expensive and characterized by a strong egalitarian universalism, raised significant problems, which eventually stopped the project half-way.

Devolution in Italy had been following a tortuous and accidental path extended over more than fifty years. In the absence of salient territorial divisions based on religion, ethnicity, or language, like Quebec in Canada, or rooted in historical tradition, as with the Celtic nations in the United Kingdom, the regionalization of Italy has tended to be viewed essentially as a political process accompanied by intense ideological polarization (Hine 1996; Roux 2008) a "story of partisan logic" (Mazzoleni 2009). This particularly happened with health and healthcare caught between the state, the regions, and the local authority, the commune. As we have seen, in most of its history since unification it is the commune, and not the regions, which in Italy constitutes the centre of the administration of healthcare. The communes are the "natural" identity level of the territories. Historically, some regional boundaries derive from *compartimenti* (compartments),which were entities artificially created for statistical and administrative purposes through the aggregation of distinct smaller units and located between the communes and the central state (on compartments as a mode of representing Italy in statistical studies and social surveys see Patriarca 1996, p. 207). The question of the design of state-region relations cannot therefore be satisfactorily treated unless we also look at the role of the communes, which have often been institutional allies of the state with the common aim of protecting competences claimed by the region. Competition between the regions and communes helps explain their varying fortunes as repositories of competences devolved by the state in the field of healthcare.

The 1970s were a period of intense activity for the regions and local authorities, which quickly became hothouses for innovation and tested out new functions and organizational forms that were often incorporated in national legislation. The period between the demise of the sickness funds in 1974 and the creation of the SSN in 1978 was when the regions began to behave as the principal actors in the arena of national health policy. They took over hospital planning and financing and were closely involved in the drafting of Law 833. The apparent paradox of a national system committed to geographical uniformity and equity being actively promoted by political actors with a regional basis is explained, in part, by the desire of the regions to take over the political room left vacant by the abolition of the highly centralized sickness funds. However, the regions were frustrated in this since the national legislation clearly intended the local authority, the commune, to be the administrative fulcrum of the new arrangements. The imbalance between the general planning functions of the regions and the active role of the communes in the governance of the USLs was a major source of tension in the arrangements for multilevel governance devised for the SSN. This was the product of an institutional alliance between the national government, anxious about the threat posed by the regions to its hegemony, and the communes, wary of the new institutional actor endowed with control functions (Fargion 1997, p. 101).

The 1992–1993 "reform of the health care reform" sparked off a competitive process of region building which gave the regions a significant role in the national political arena vis-à-vis the central state. Regionalization of health governance officially invoked the standard "democratic" merits of federalism, in particular the greater proximity of government to the preferences of local citizens. In fact, the main purpose of the reform was to increase the financial accountability of the regions by requiring them to cover their deficits with an increase in own-source taxes and patient co-payments. Given the regions' limited fiscal capacity, their commitment to fund their often large deficits with the meagre own-source revenues was scarcely credible. However, this reallocation of responsibility in healthcare galvanized the regions into an intense process of competitive region building aimed at strengthening their political identity and visibility and subtracting policy space from the central state (Banting 1995). Competition among the regions and between the regions and the state continued for most of the decade. Initially, the principal issue at hand concerned the choice of the organizational model for the regional health services. Later in the debate, the question of intergovernmental relations in the development of national health policies took centre stage. In this first phase of the regionalization of the SSN, the chief loser was local government, which national legislation stripped of all administrative responsibilities over the ASLs.

At the end of the 1990s, pressure for greater regional autonomy came from different quarters. The political field was marked by the overwhelming success of the *Lega Nord* (Northern League), the first Italian party to support federalist or openly secessionist demands outside the local protection of linguistic or ethnic minorities (Giordano 2000). Healthcare was the leading sector in the disorderly process of devolution of powers. It had become, by far, the major social function devolved to the regions as education, labour policy, and pensions in Italy are the responsibility of the state. Health expenditures were the largest item in the regional budgets, on average about 70 per cent of their total expenditure. It was also a salient political issue, as the perceived quality of the regional health services significantly influences citizens' views of the regional government. As a result, the directly elected "Governors" (as the Presidents like to be called) of the regions are acutely sensitive about public opinion over regional health service. The northern regions were the main actors. They had acquired power *de facto* in structuring the regional system according to their political and ideological preferences and negotiated national policies extolling their veto power. They frequently used recourse to the Constitutional Court contesting the right of the central government to intervene directly in the organization and management of the regional health services (Balduzzi 2004).

Pressure from the top came from the Treasury, whose policies were increasingly focused on containing public expenditures. The cuts in the SSN budget had significantly contributed to entering the European Monetary Union in the previous decade (Mapelli 2000), and public health expenditures were still relatively low by international standards. In 2002, public health expenditures were 6.4 per cent of GDP, while total health expenditures were 8.6 per cent of GDP (OECD 2003). Beginning in 1999, Italy became a member of the European Monetary Union with rigid budget and debt criteria. However, Italy's public debt as a proportion of GDP stood at over 100 per cent compared with the European Monetary Union requirement of no more than 60 per cent, and this made public debt a structural factor in Italian fiscal policy. Health policymaking operated in the shadow cast by the towering public debt, by far the largest in the European Union. This explains why, despite relatively low public health spending, there were continual calls for further cutbacks from, for example, the Treasury, *Banca d'Italia*, the national central bank, the International Monetary Fund, and of course, the European Union. Public healthcare was the standard target for spending reviews. In addition, devolving fiscal responsibility permitted the central government to shift to the regions the blame for expenditure overruns as well as the political costs of cost containment, including delisting of covered drugs, services, and co-payments.

Constitutional reform recognized the sovereignty of the regions in several areas and listed the powers, which are the exclusive competence of the state, and those for which the state and the regions have concurrent responsibility. The amended Constitution provided regional governments concurrent or exclusive legislative powers including, among others, healthcare, education, scientific research, land use, energy production, distribution, and transport. The constitutional foundation of the healthcare system was based on a joint jurisdiction, with the regions holding the legislative powers concerning healthcare within the general principles set out by the central government. However, the state had exclusive competence in "the determination of the essential levels of services concerning civil and social rights", including the basic package of health services to be uniformly guaranteed to all by the SSN (Balduzzi 2002). In particular, the regions had *potestà concorrente* (shared authority) over the general function of *tutela della salute* (protection of health) much broader (and ambiguous) than the original and specific matter of "health services and hospitals" of the 1948 Constitution. This was taken to imply that the regions had virtually complete responsibility for the organization and administration of the SSN, except for negotiating SSN staff working and career conditions, remuneration levels of staff and contracted providers, and prescription drug prices. The expansive nature of the "health protection" subject tends to intertwine with various other competences such as environmental protection, food security, safety at work, and scientific research. The new division of responsibility and authority therefore combined the traditional "vertical" problems relating to the demarcation of responsibility between levels of government, with new and original problems of a "horizontal" nature deriving from the uncertain boundaries of the matters of competence. This contributed to fuelling the dispute between the state and the regions observed in the early years of the constitutional reform.

The 2001 Constitutional reform aimed at a classic model of cooperative federalism (France 2008). A plan for fiscal federalism promised the regions more revenue and spending autonomy to perform their additional responsibilities. Revenues were to come both from own-source taxes and from revenue sharing of central government taxes, plus transfers from a national equalization fund administered by the state. No conditions were to be placed on the use of state transfers, a safeguard against abuse by the state of the vertical fiscal imbalance characterizing state-region fiscal relations. A ban on borrowing to finance current expenditures aimed at tightening the budget constraints (Giarda 2001). These intensions were translated into a detailed plan for "fiscal federalism", which applied to all sectors but had health as the principal target. The goal was to introduce virtually complete fiscal decentralization over a period of thirteen years. A business tax (IRAP) was to be the major source of regional revenues, accounting for some 74 per cent of the total. This tax on public and private productive activities has an extremely large base, approximating 50 per cent of the GDP, but is also the most volatile and unequally distributed across the regions. For example, the per capita revenue of the private part of the tax base (over which the regions exert their autonomy) in 1998 was 1,208,700 lire in Lombardy and 180,400 lire in Calabria. The second most important tax was a regional surcharge on the individual income tax (IRPEF), providing an estimated 6.4 per cent of regional revenues. The regions were allowed to vary the rates for the business tax and the personal income tax surcharge within a band set by the central government (Arachi and Zanardi 2000).

The most important problem for the stability of fiscal federalism was, and still is, the persistent dualism of the Italian economy, and by consequence, the wide differences in the fiscal capacity of the regions, particularly along the north-south divide. The International Monetary Fund highlighted major imbalances between own-source revenue and current expenditure obligations, which varied significantly among the regions (IMF 2000). Five regions in the centre-north had surpluses, while all southern regions had significant deficits. Calabria own-source revenues, for example, covered only 30 per cent of its total public expenditures. Imbalances between fiscal capacity and regional public expenditures were especially relevant for health expenditures, which are the largest part of the regional budget and tend to grow at a faster rate than GDP. This meant that, even if the regional tax system could capture the entire economic growth of a region, regional own-source revenues would inevitably be swallowed up by the growth in health expenditure. A vertical equalization fund had to ensure that all regions have the resources to uniformly provide the services defined by LEA. The fund was financed by the VAT revenues (which are more uniformly distributed than IRAP and IRPEF) ceded by the central government. Distribution to the regions was to be based on an allocation formula adjusted for fiscal capacity (but limited to 90 per cent, to stimulate growth) and the size of each region, to account for economies of scope in the largest regions. The scheme was to be phased in over a period of twelve years. The regions, after three years during which they would be required to spend what the central government calculates is necessary to ensure the LEA, would be free to decide how much to spend on health. Failure to guarantee the LEA would be sanctioned by a pre-specified reduction in the equalization transfer and other transfers.

The national equalization fund to help the poorer regions meet the costs of the devolved responsibilities reflects the declared commitment to a solidaristic and cooperative form of federalism. However, continued dependence on grant financing via the equalization fund perpetuates the separation of spending and financing decisions, which is the root cause of inadequate respect for budget constraints (France 2005). This invites opportunistic behaviour by the regions and leads to conflict with the central government over the adequacy of central funding to meet spending needs. To reduce this risk, a limit on the duration of the equalization fund was set, and identification of the donor and beneficiary regions was considered. However, it is doubtful that a dozen years was time enough to make any significant progress in reducing the long-standing north-south economic divide. Moreover, naming the donor and beneficiary regions and making the redistribution process more transparent risked creating tensions in the nascent federation and reduced the political reputation and eventually the autonomy, of the financially dependent regions. Perceptions that transfers were being used inefficiently, for example, risked fuelling secessionist sentiments in the donor regions, for example, by lowering their fiscal effort, which is not easily observable by the central government and the other regions. Moreover, the less developed regions were caught in a sort of "poverty trap". On the one hand, they would prefer high tax rates, in order to increase revenues and match expenditure responsibilities. They may want instead low tax rates in order to attract business and promote economic growth, accepting a short-term reduction in their own revenues in the hope of enlarging their tax base in the long term.

The LEA, the explicit package of services to be guaranteed to all Italian citizens, served several functions in the devolution of powers, primarily as a national standard to keep the regions accountable with their citizens and the state and to measure the resources needed by the SSN and its regional segmentations (Torbica and Fattore 2005). Law based the LEA on four criteria: human dignity, clinical effectiveness, appropriateness, and technical efficiency (Taroni 2001). A positive list very broadly defined the services that the SSN was actually providing, in its three main areas of public health, community, and hospital care. A negative list more specifically included three categories of services that should no longer be provided on the public purse. A few, miscellaneous services were outright excluded from coverage, because of their proven clinical inefficacy or because they were not deemed pertinent to the goals and aims of the SSN, such as cosmetic surgery. A list of ambulatory specialist services was to be excluded from public coverage on a case-by-case basis, considering their clinical or organizational appropriateness for individual clinical indications. The last category of services in the negative list included an indicative list of potentially inappropriate hospital admissions, which the regions were required to move to alternative levels of care. Examples include admissions for carpal tunnel release, uncomplicated hypertension, and diabetes.

The very pragmatic approach adopted summarized the four meanings of the concept of medical necessity that had emerged in the past as well as current medical debates about the appropriate level of service coverage, including the Oregon Plan described previously (see page 219). The principal issues included "what doctors and hospitals currently do", "what is consistently funded across all provinces (or regions, in Italy)", "the maximum we can afford", and "what is scientifically justified" (Charles et al. 1997). Both the soundness of the conceptual approach and the effectiveness of the policy are open to debate. The notion of an explicit list of services to be included in the basic healthcare package had proved to be technically difficult to define and politically and legally next to impossible to enforce (see, e.g., Coulter and Ham 2000). The decision to define the national character of the SSN solely in terms of the LEA, and their use as an instrument of inter-institutional reciprocal accountability, proved ineffective. A multi-dimensional approach may have been more appropriate, similar to the 1984 Canada Health Act, which requires that the provinces to be eligible for federal funds for health must guarantee comprehensiveness (deliberately left vague) but also portability across provinces, accessibility, universality, and public administration (Flood and Choudry 2002). The very short negative list of the Italian LEA certainly defeated the expectations of those hoping that delisting covered services would make a significant dent on public expenditure and provide space for the development of complementary mutual funds. Moreover, the concept of a negative list of services to be excluded from current provision made the LEA a difficult tool for calculating the central government contribution to the SSN.

The governance of the disjointed form of "marble-cake" federalism, which was emerging at the dawn of the century, would have required a robust system of institutions facilitating the relations among the regions and between the regions and the state. However, the amended Constitution did not provide for the direct participation of the regions in national policymaking. Both the regions and the state can request ex post judicial review by the Constitutional Court if they consider that their powers have been infringed. Appeals to the Constitutional Court by both levels of government, but especially by the state, increased significantly from 25 in 2000 to 95 in 2002, reaching a peak of 115 in 2004 (Ronchetti 2008). Conflict over the interpretation of budget constraints was an important reason for appeals, and the healthcare sector was one of the principal arenas of dispute. Apparently, for a short period the Court was the arbiter in the allocation of responsibilities. Coordination and the difficult balance between the countervailing powers of the regions and the state remained entrusted to the so-called System of the Conferences, including State-Regions, State-Local Autonomies, and the Unified Conference (Carpani 2006). The System of the Conferences was defined by the Constitutional Court in 1994 as "an institution operating within the national community as an instrument for the implementation of cooperation among them" with the task of "realizing loyal collaboration and promoting agreements and understandings". "Agreements" allowed the regions to participate indirectly in national legislative activity through the commitments undertaken by the central government when meeting in the Conferences. These are, however, just political agreements. The

Constitutional Court, in rejecting the suit of several regions against the national government for its alleged failure to respect accords reached in Conference, denied the possibility that such accords could limit the Parliament's legislative powers.

A change in government did not substantially change the policies of devolution. The new centre-right government broadened existing policy explicitly linking devolution, liberalization, and privatization. Policies of vertical devolution of powers from the central government to the regions were accompanied by plans for a "horizontal" transfer of functions to the market and the so-called third sector, including families, voluntary organizations, foundations, mutual funds, and insurance companies. The new welfare-mix and its community welfare aimed at creating a "multi-pillar" system on the model of the pension reform of the early 1990s, where the provision of a bare minimum of services funded by general taxation would be complemented by mutual funds (the second pillar) and health insurance companies (the third pillar), given appropriate tax incentives. The hybrid system would accomplish a variety of objectives: attract additional private resources, expand choice, and increase efficiency and innovation through competition in the supply and demand sides of services (Micossi and Paoli 2002). Later, the white paper "The good life in the active society" sought a "new social model" "not only concerning public functions but also recognizing the value of the family, of not-for-profit and for-profit enterprises, as well as of all the intermediate bodies that contribute to forming a community" (Ministero del Lavoro 2009, p. 2). The appeal to the "active society" is reminiscent of the Britain conservative Prime Minister David Cameron's later call for a "Big Society" to take over the provision of public services and shrink the role of the state from provider to "enabler". This classic form of market-making federalism (Bennet 1990) however never materialized and remained a vague political aim.

Contractual Relations: Patti per la Salute and Piani di Rientro

While debates over "federalism" were prominent in the political theatres, in the real world the regions and the central government played quite a different game. The way policy was developed and where it took place had changed. The venue for the development of health services had shifted to the regional level, and the regions had become powerful veto points in the formulation of national policies. National health policy was now made less in Parliament and in the central ministries and more through negotiation between the national government and the regions, something akin to executive or cabinet federalism familiar to Canadians. Its content was more focused on fiscal policies than on promoting health and planning healthcare. The intense debate over the future fiscal federalism notwithstanding, the vertical fiscal imbalance between regions and the state had remained. Own-source regional revenues had become increasingly inadequate to meet the financial needs created by the devolution of powers. A major downward shift of responsibility for organizing and operating health services would ideally be matched by devolution of resources and responsibility for financing. Failure to reform the fiscal system to match the devolution of powers had inevitably led to the growth of central spending power. The root problem was the skewed nature of intergovernmental fiscal relations: the regions had gained ever-increasing spending power in healthcare but limited responsibility for raising the revenue to finance this, allegedly causing a serious accountability problem. This resulted in a new form of the traditional divide between spending power and public financing. Moreover, health policymaking operated in the shadow cast by the towering public debt, still by far the largest in the European Union. The imperative of containing public expenditure dominated the political agenda concerning the public health sector, which was the second largest source of public expenditure, estimated at about 13 per cent of total expenditure (Istat 2009). This made the SSN the main target for reducing public expenditure, even if it was not expensive in relative terms.

In 2007 public health spending in Italy, at 2686 US dollars in purchasing power parity (PPP), was well below the OECD average of $PPP 2964. Between 2000 and 2007 health spending per capita increased in real terms by 1.9 per cent per year compared to 3.7 per cent in OECD countries (OECD 2007, 2009). As the power of the purse was firmly in the hands of the central government, the Treasury annually determined the "requirement" (*fabbisogno*) of the SSN and resurrected the old strategy of underfunding to keep the regions in a state of permanent fiscal crisis. For their part, regional health services confirmed their history of overshooting planned spending. Between 2001 and 2005, the regions accumulated debt amounting to almost 21 billion euro. Initially deficits were evenly spread, but within a few years major deficits concentrated in a few regions: in 2007, three regions (Lazio, Campania, and Sicily) accounted for over 82 per cent of the global deficit (Tediosi et al. 2009). The main terms of the health policy agenda were the tight conditions attached to central transfers and the strict procedures to check regional compliance. Health policy concentrated on budgetary policy negotiated with the Treasury. In effect, the intergovernmental accords signed beginning in 2000 sanctioned a significant increase in the exercise of the central spending power, which was used principally to steer the health policy of the regions.

The first indications that the national government intended to go down this road were contained in the 1998 Internal Fiscal Stability Pact, which translated the EU Stability Pact at the domestic level. It made regional budgetary balance a condition for receipt of a reserved quota of 1.5 per cent of the National Health Fund, now renamed "interregional" in formal obeisance to federalist principles. This strategy was refined in the *Patto per la salute* (Health Pact) of 2000 negotiated with the centre-left government, which set new criteria for allocating the burden of eliminating the deficit between the state and the regions. The Pact also committed the regions to taking specific actions to eliminate their share of the deficit. The strategy was further honed in the 2001 Pact, which committed the centre-right government to bringing public health spending up to 6 per cent of GDP over three years and to defining the LEA entitlement and estimating its cost. The government also worked towards obtaining rapid parliamentary approval for a series of measures requested by the regions to help them to reach budget balance. The regions promised to take action to limit spending on prescribed drugs to 13 per cent of total expenditure, increase fiscal effort, exclude certain drugs from SSN coverage (so-called delisting), and reduce bed capacity to four beds per 1000 inhabitants. A 2005 accord went into greater detail, making access to increased funding by the single regions (funding, i.e., which had already been appropriated for financing the SSN) conditional on their adopting measures of administration, for example, setting up purchasing unions and meeting reporting requirements for expenditure monitoring. The sanction for non-compliance was denied access to a so-called supplementary fund, which was, in reality, a portion of funding already earmarked for the SSN.

The Budget Balance Plan (*Piani di Rientro*) strategy, first adopted in 2006, may in a sense be considered yet another attempt to force tight budget constraints on the regions, using a mix of conditionality and financial and political sanctions. While *Patti per la Salute* contain general rules applicable to all regions. Budget Balance Plans are the product of bilateral negotiations between the state and a single region which was officially declared by the central government to be "in financial difficulty" in terms of the size of its deficit, defined as over 7 per cent of its budget, later reduced to 5 per cent. Additional criteria, such as chronicity of the deficit and the size of the debt, were added over the years to encompass half of the Italian regions. In case of large recurring deficits and high debt level a *Commissario* (provisional administrator) appointed by the Treasury took over responsibility for health administration from the President of the Region. The policy initially focused on the seven regions where the global SSN deficit was concentrated. Each of these regions was paired with a "mentor" (non-deficit) region selected by the Conference and had to submit to quarterly monitoring by the central authorities, led by the Treasury. The problem regions, by co-signing the plan along with the Ministry of the Economy, undertook to balance their budgets (current expenditures) within a period of three years (renewable) and to pay off their debt accumulated through 2005. Implementation of the measures were to be verified before access was allowed to the central financing in question, including a long-term loan with favourable terms guaranteed by the state (Tediosi and Paradiso 2008). Results were astounding in financial terms. Regions under *Piani di rientro* reduced their deficits from 4.7 billion euros in 2006 to less than 400 million in 2018. The total SSN deficit declined from 6.5 per cent of its budget in 2006 to 2.5 per cent in 2015 and 1.1 per cent in 2018 (Ufficio Parlamentare di Bilancio 2019). Analyses suggest that the focus on short-term financial objectives has encouraged the regions to opt for measures with rapid but short-lived results (Tediosi et al. 2009). In addition to increases in the tax burden through the additional IRAP and IRPEF and across-the-board hiring freezes, the regions resorted, in the main, to increase co-payments for drugs and services, reduce tariffs paid to contracted providers, and organize centralized purchasing and distribution to contain drug expenditures (Pavolini 2012).

The accords and pacts between the state and regions are a kind of "shared-cost federalism", as opposed to "classical" and "joint decision" federalism (Banting 2007) in which the national government grants financial support to regional governments on precise jointly negotiated conditions. Shared-cost programmes mean however incomplete contracting and create ample possibilities for opportunistic behaviour by both parties, for blame shifting and, therefore, for conflict (Banting 2007, p. 152). The main difference between Canada and Italy is that in the latter case the funding in question is not new money for additional programmes but is subtracted from the sum of resources already appropriated to finance the SSN. A region has therefore little option but to accept the conditional grant.

The most important element of national health policies of this period was their progressive absorption in the context of budgetary policies dominated by the Treasury both towards the regions and with the Ministry of Health. The subordination of health policies to public finance objectives countered their formal transfer to the regions. Regions saw their organizational powers, which had been formally increased, diminished in practice by their persistent lack of financial autonomy. The central government had maintained responsibility for the *ex ante* financing and has reserved ever-wider discretions for *ex post* intervention on the deficits that it helped to create with its underfunding. The state was exploiting the financial difficulties of the regions and using its spending power to increase its influence on regional health policy. Health policies designed at the national level are mostly to be seen as organizational corollaries to the financial pacts, with periodic re-negotiation of the array of measures necessary to meet global and specific spending ceilings and capacity limits, which then become conditions for the receipt of state transfers. The immediate effect of this strategy, which had been followed beginning in 2000 regardless of the government in power, was to focus attention on the purely financial aspects of health policy, downgrading questions like service organization, health promotion and prevention, and inequalities. These were in obvious contrast with the continuing federalist discourse. Whether the contractual evolution in the relations between the state and the regions was the expression of the new wave of re-centralizing policies, which had started in many health systems, such as the Scandinavian countries, where decentralization policies have started earlier and proceeded more intensely (Saltman 2008). If they would have gone further is impossible to say: the Global Financial Crisis and then the COVID-19 pandemic changed everything.

References

Arachi, G., Zanardi, A. (2000) La riforma del finanziamento delle regioni italiane: problemi e proposte. Economia pubblica, 30 (1), pp. 6–43

Balduzzi, R. (2002) Titolo V e tutela della salute. Quaderni Regionali, 1, pp. 65–85

Balduzzi, R. (2004) La creazione di nuovi modelli sanitari regionali e il ruolo della Conferenza Stato-Regioni (con una digressione sull'attuazione concreta del principio di sussidiarietà "orizzontale"). Quaderni Regionali, 3, pp. 11–26

Banting, K.G. (1995) The welfare state as statecraft: Territorial politics and Canadian Social Policy In: European social policy: between fragmentation and integration. Eds. S. Leibfried, P. Pierson. Washington D.C., Brooking Institution, pp. 269–300

Banting, K.G. (2007) The three federalisms: Social policy and intergovernmental decision-making. In Canadian federalism: Performance, effectiveness and legislation. Eds. H. Bakvis, D. Skogstad. Oxford, Oxford University Press, pp. 137–160

Bennet, R.J., Ed. (1990) Decentralization, local governments and market: Towards post-welfare agenda. Oxford, Clarendon Press

Buglione, E., France, G. (1983) Skewed fiscal federalism in Italy: implications for public expenditure control. Public Budgeting and Finance, 3 (3), pp. 43–63

Carpani, G. (2006) La Conferenza Stato-Regioni. Competenze e modalità di funzionamento dall'istituzione ad oggi. Bologna, Il Mulino

Charles, C., Lomas, J., Bhatia, M.V., Vincent, V.A. (1997) Medical necessity in Canadian health policy: four meanings and... a funeral? Milbank Quarterly, 75 (3), pp. 365–394

Coulter, A., Ham, C., Eds. (2000) The global challenge of health care rationing. Buckingham, Open University Press, pp. 15–26

Fargion, V. (1997) Geografia della cittadinanza sociale in Italia. Bologna, Il Mulino

Flood, C.M., Choudry, S. (2002) Strengthening the foundations: Modernizing the Canada Health Act. Discussion Paper nr.13. Commission on the Future of Health Care in Canada

France, G. (2005) Diritto alla salute, devolution e contenimento della spesa: scelte difficili, scelte obbligate. In: Terzo rapporto annuale sullo stato del regionalismo in Italia. ISSIRFA Milano, Giuffrè, pp. 231–247

France, G. (2008) The form and context of federalism: Meanings for health care financing. Journal of Health Politics, Policy and Law, 33, pp. 649–705

Giarda, P. (2001) Le regole del federalismo fiscale nell'articolo 119: Un economista di fronte alla nuova Costituzione. Le Regioni, 29 (6), pp. 1425–1483

Giordano, B. (2000) Italian regionalism or "Padanian" nationalism—the political project of the Lega Nord in Italian politics. Political Geography, 19, pp. 445–471

Hine, D. (1996) Federalism, regionalism and unitary state: Contemporary regional pressures in historical perspective. In: Italian regionalism: History, identity and politics. Ed. C. Levy. Oxford, Berg, pp. 109–130

IMF—International Monetary Fund (2000) Italy: selected issues. IMF Staff Country Report No. 00/82. IMF, Washington D.C.

ISTAT (2009) Spesa delle amministrazioni pubbliche per funzione: Anni 1990–2007. Roma, Istat

Mapelli, V. (2000) Tre riforme sanitarie e ventuno modelli regionali. In: L. Bernardi Ed. La finanza pubblica italiana: Rapporto 2000. Bologna, Il Mulino, pp. 299–327

Mazzoleni, M. (2009) The Italian regionalization: a story of partisan logics. Modern Italy, 14 (2), pp. 135–150

Micossi, P., Paoli, G. (2002) Il cittadino e le garanzie offerte dal SSN: livelli di assistenza e finanziamento. In: N. Falcitelli, M. Trabucchi, F. Vanara (a cura di) Rapporto sanità 2002: Il cittadino ed il futuro del Servizio sanitario nazionale. Bologna, Il Mulino, pp. 95–132

Ministero del Lavoro, della Salute e delle Politiche Sociali (2009) Libro Bianco sul futuro del modello sociale. La vita buona in una società attiva. Roma, Ministero del Lavoro

OECD—Organization for Economic Cooperation and Development (2003) OECD Health data 2003. Paris,OECD

OECD—Organization for Economic Co-operation and Development (2007) OECD Economic Surveys: Italy. 2007/12, Paris, OECD

OECD—Organization for Economic Co-operation and development (2009) OECD Health Data 2009: Statistics and indicators for 30 countries. Paris, OECD

Patriarca, S. (1996) Numbers and Nationhood: Writing statistics in nineteenth-century Italy. Cambridge, Cambridge University Press

Pavolini, E., Ed. (2012) Il cambiamento possibile: La sanità in Italia tra Nord e Sud. Roma, Donzelli

Ronchetti, L. (2008) Il contenzioso e la giurisprudenza costituzionale nel regionalismo italiano dopo il 2001. In: Quinto rapporto annuale sullo stato del regionalismo in Italia. Milano, Giuffrè, pp. 121–126

Roux, C. (2008) Italy's path to federalism: Origins and paradoxes. Journal of Modern Italian Studies, 13 (3), pp. 325–339

Saltman, R.B. (2008) Decentralization, re-centralization and future European health policy. European Journal of Public Health, 18 (2), pp. 104–106

Taroni, F. (2001) Livelli essenziali di assistenza: sogno, miraggio o nemesi? In: Fiorentini G. Ed. I servizi sanitari in Italia,2000. Bologna, Il Mulino, pp. 27–91

Tediosi, F. Paradiso, M. (2008) Il controllo della spesa sanitaria e la credibilità dei piani di rientro. Rapporto ISAE, Roma, pp. 125–144

Tediosi, F. Gabriele, S., Longo, F. (2009) Governing decentralization in health care under tough budget constraint. What can we learn from the Italian experience? Health Policy, 90, pp. 303–312

Torbica, A., Fattore, G. (2005) The Essential levels of Care in Italy: when being explicit serves the devolution of powers. European Journal of Health Economics, 6, pp. 46–52

Ufficio Parlamentare di Bilancio (2019) Lo stato della sanità in Italia. Focus Tematico n° 6

CHAPTER 11

Two Converging Crises

The early years of the twenty-first century witnessed disjointed changes in health policy: a constitutional devolution of power to the regions and, conversely, de facto fiscal governance that was weighted in favour of the central government's fiscal objectives. This changed the direction of the health policy debate, which focused on its fiscal dimensions, as it was generally accepted that organizational issues were a regional responsibility. The new and tighter fiscal policy showed remarkable continuity between governments of the centre-right and the centre-left. It reflected the strengthening of the European Union fiscal policy through its Stability and Growth Pact, which had been translated into domestic policy as the Internal Stability Pact. The stagnant political climate was shattered by the two converging crises of the Great Recession and the COVID-19 pandemic.

In Italy, the Global Financial Crisis of 2008 began in earnest in the summer of 2011 with a dramatic increase in the bond spread with Germany. A letter from the European Central Bank (ECB) asked Italy to undertake specific reforms in exchange for help in getting relief on its sovereign debt that was under attack. This opened the way to an emergency government of technocrats led by Mario Monti, a university professor of economics and former European Commissioner. This mirrored the institutional response to the economic crisis of the early 1990s. However, instead of offering an external constraint, the European Commission and the ECB now explicitly claimed authority over domestic policies. While important

F. Taroni, *Health and Healthcare Policy in Italy since 1861*,
https://doi.org/10.1007/978-3-030-88731-5_11

reforms of the pension and the labour market were passed, health policy changes simply strengthened and accelerated processes and strategies already in action before the outbreak of the crisis. The SSN budget continued to be the focus for cutting public spending. Co-payments were increased with what came to be known as the "super-ticket". Legislative attempts at structural reform of health service organization faced the difficulties of intervening in a field, which had firmly been appropriated by the regions and allowed for only broad goals and vague directives of the soft law epitomized by Patti per la salute (pacts for health), negotiated between the regions and the central government. The slashing of the SSN budget continued over the several years of austerity politics, focusing on the freeze in hiring and pay of health professionals. Moreover, the politics and policies of the austerity years resurrected the tensions between universalism and selectivity and fuelled secessionist movements in the wealthiest regions of Veneto, Lombardy, and Emilia-Romagna, which collectively encompass about 40 per cent of Italian GDP. This all changed again with the outbreak of the COVID-19 pandemic in February–March 2020. The outbreak of the pandemic found an SSN exhausted by over ten years of austerity politics, barely able to keep up with the surge of infections.

Weathering the Great Recession

Between 2001 and 2010, Italy experienced the lowest average per capita growth rate, the longest recession, and the weakest recovery of the Eurozone, the countries that share Europe's single currency. Over the decade, Italy had often been characterized as "the sick man of Europe", a vulnerable economy, and a risk to the EU's financial stability. Moreover, the increase in state capacity that emerged in the 1990s seemed transitory. The capacity to learn from others and from experience to make informed choices over policy instruments and their settings had reflected more a reliance on outside experts than on the capacity of permanent civil servants. The empowerment of the executive had been made possible by the contingent suspension of ordinary parliamentary activity. As the contingent suspension was lifted, traditional party governance resumed. The process of policymaking was frantic and disjointed, with difficult relations between central governments and the regions. This was the worst of conditions to weather the Global Financial Crisis, which hit the country in the summer of 2011, when the Italian sovereign debt was under attack in the financial markets. Between May and July 2011, the yield spread with

Germany increased by 100 basis points, from 170 to 270, then reached 400 basis points in September, and surpassed 500 basis points in November. The Italian government was slow in responding, and for a few months, it let people believe that the unsophisticated configuration of financial activities in Italy would have spared the country from crisis (Quaglia 2009). In fact, the crisis cost Italy over 10 per cent of its GDP, which decreased to the levels of the mid-1990s. The general unemployment rate redoubled to 11.9 per cent. Among the youngest, unemployment rose from less than 20 per cent to over 38 and then to a staggering 50 per cent. The IMF expected that the Italian economy would not return to the levels seen before the crisis until mid-2020. Unfortunately, the outbreak of the COVID-19 pandemic in the early months of that year did not allow the expected recovery to materialize.

The crisis began in the summer of 2011 as contagion from the Greek economic crisis and precipitated with a letter from the ECB listing the reforms and adjustments that Italy had to pass in order to get relief on its sovereign debt. Italy was not subject to the Economic Adjustment Program (EAP), also known as Memoranda of Understanding, which were managed by the EU, ECB, and IMF (the so-called Troika), as were Ireland, Portugal, and Greece. It was nevertheless under strict scrutiny and specific pressures from the outside. The ECB letter, as published by Corriere della Sera on September 29, 2011, urged Prime Minister Silvio Berlusconi "to take immediate and bold measures to ensure the sustainability of public finances" and explicitly mentioned seniority pension and labour market reforms. The re-emergence of the external constraint of European institutions over Italian policies was the first similarity with the crisis of the early 1990s. The second was the suspension of normal politics and the formation of a technocratic executive with the specific task of weathering the acute financial crisis (Marangoni 2012). When Mario Monti became prime minister, the government was completely staffed by outsiders from the political world, drawn from academia, banking, business, and the highest echelons of the civil service. The emergency government was supported by an extremely large majority slightly short of 90 per cent of the votes in both chambers of the Parliament, which however was anything but politically homogeneous. The recurrence of a "technical" government was reminiscent of the Ciampi and Dini governments of the mid-1990s and represented the failure of the party system to consolidate after its meltdown of twenty years earlier (Bull and Pasquino 2018).

Bill Emmott pointed to the analogies between the current sovereign debt crisis and the economic and institutional crisis of 1992–1993, suggesting that this was "Italy's second chance" to "truly rebuild and reform either its politics or its economy after a nasty shock" based on previous experiences (Emmott 2012, pp. 1–2). However, the two crises also had several differences. While the 1992–1993 crisis was essentially a domestic problem limited to Italy, the current crisis was a risk for all countries of the Eurozone. Measured by the ratio of public debt to GDP, Italy was the second largest debtor of the Eurozone, and Greece was the first with over 150 per cent. However, while Greece accounts for only about 2 per cent of the Eurozone GDP, Italy accounts for over 13 per cent. Therefore, the Italian public debt was not just a large burden for the country but was also "big enough" to be a serious threat for the stability of the entire Eurozone. Another significant difference, strictly intertwined with the former, was that the European political and financial institutions were claiming explicit authority over domestic policies of the member states. This was "the third face" of the EU that moved from being an exogenous actor to become an enduring stakeholder in domestic policies as discussed above (Greer 2014).

According to Greer, the first two phases initially included networking and collaboration between member states and then a regulatory phase promoting the "four freedoms" of the integrated European market, the free movement of goods, services, capital, and people. The exceptional circumstances of the great recession prompted the EU to impose strict and detailed oversight of member states' budgets and policies in pursuit of fiscal rigour. The EU interventionist policy was a factor in the subtle change of disposition towards the EU observed in Italy throughout the decade. Bull and Pasquino have argued that the positive disposition towards the European constraints meant that Italy needed the help and discipline of the EU but implied also a more general attitude that "travelling the European road" would bring many good things (Bull and Pasquino 2018). The imposing presence of the EU, particularly through the adoption of the politics of austerity, undermined that positive attitude and gave rise to a diffuse sense of scepticism in the population and a more challenging approach to the EU in several political parties, and also in some governments of the time.

The ECB letter's recommendations included both short-term and structural adjustments to entrench fiscal rigour but did not mention the health sector. Detailed prescriptions for health system reforms were part of the national bailouts for Ireland and Greece in 2010, which had to accept both a cap on public health expenditure set at 6 per cent of the GDP and a series of microeconomics measures on drug prices, use of generics, staff salaries, and so on (Economou et al. 2014). The programme for Portugal, for example, included thirty-three commitments widely affecting the organization and financing of health services. In Italy, the government's spending review in the health field consisted essentially in accelerating and strengthening the fiscal policies already in progress before the outbreak of the crisis. These aimed to slash health spending to rein in public budgets, as health expenditures in Italy are funded three-fourths from public sources and represent the second largest item of public spending, at about 13 per cent of the total. As with the crisis of the early 1990s, during the 2008 crisis, and the subsequent period of austerity politics, the SSN assumed the traditional role of "lender of first resort", the treasure island for the cutbacks of public spending, combining its large size with the guarantee of prompt returns and relatively low visibility.

From 2008 to 2015 and beyond, public health expenditures remained stable at about 6.6 per cent of a shrinking GDP, according to the WHO Global health expenditure database (Fig. 11.1). The average rate of growth declined from 7.1 per cent in the period 2001–2006 to slightly less than 3 per cent in the period 2007–2010, and then fell to zero in the crisis period 2010–2015, when for three consecutive years (2011, 2012, and 2013) health expenditure decreased in absolute terms (Ministero

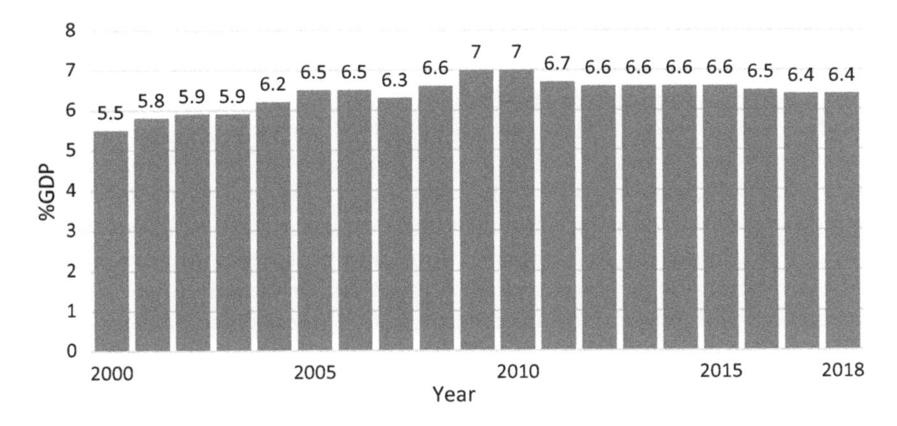

Fig. 11.1 Public health expenditure, per cent of GDP, Italy 2000–2018. (*Source*: WHO Global Health Expenditure database)

della Economia e delle Finanze 2016). A modest average increase by 1 per cent per year was observed between 2015 and 2018, still below the pre-crisis years (Ufficio Parlamentare del Bilancio 2019).

Cutbacks in public health spending have typically been made across the board, but some sectors were hit faster and harder than others were. Much of the reduction in the SSN budget was concentrated in the category "wages and salaries of employees", which fell from 39.8 per cent in 2000 to 31.3 per cent of the declining total public expenditures in 2015 (Ministero della Economia 2016). The average rate of increase of the cost of personnel of the SSN was reduced by more than half (2.1 per cent between 2006 and 2010 compared with 4.7 per cent between 2001 and 2005) and then fell into the negative, at -1.5 per cent, in the midst of the crisis, between 2011 and 2015. Measures adopted included slowing the recruitment and the promotion of doctors and allied health professions, blocking their turnover, and freezing salaries for over six years. These resulted in the reduction of the number of personnel employed by the SSN by 6 per cent in 2017 compared to 2008 and the ageing of the work-force. The mean age of people employed by the SSN increased from 43.5 years in 2001, to 47.2 in 2009 and 50.7 years in 2017 (Ufficio Parlamentare del Bilancio 2019). Between 2005 and 2016 the proportion of physicians aged fifty-five years or less shrunk by close to 60 per cent, while those aged over fifty-five years increased by about 40 per cent (Petmesidou et al. 2020). Fifty-four per cent of physicians were aged fifty-five years or older (Vicarelli and Pavolini 2015).

SSN expenditures for purchasing contracted services from private providers also declined in absolute terms compared to the pre-crisis level, although at a much slower rate. This increased its proportion of the SSN budget, which rose to 22.0 per cent in 2015 from 18.8 per cent in 2001 and 20.7 per cent in 2010. The range of measures to cut pharmaceutical spending (which amounts to slightly less than 17 per cent of the SSN budget) included introducing centralized procurement and the direct distribution of drugs from SSN organizations, and promoting the prescribing and dispensing of generics, still used much less frequently than in other countries. The most important measures strengthened the payback from pharmaceutical industries for spending in excess of ceilings on hospital and territorial drug use and reduced the prices paid by the SSN for pharmaceuticals and medical devices by imposing a discount over pharmacies and the industries. The new structure of public spending in healthcare in Italy did not match OECD analyses, which concluded that in European countries "the greatest decreases have been observed in pharmaceutical spending and in areas of public health and prevention" (Morgan and Astolfi 2015).

The aggregate effects on health expenditures seem obvious. OECD data from various years show that in Italy total health spending per capita in US dollars at purchasing power parity declined in 2013 ($ 2965) through 2017 ($ 3033) compared to the $ 3146 PPP in the pre-crisis year 2008. This increased the gap between Italy and the mean public health expenditures of the first fifteen European countries (EU15), which fell to 77.1 per cent in 2011 from 85.5 per cent in 2000 (Pavolini et al. 2015), and were at $ 3732 PPP, $ 3936, and $ 4084 respectively in the same years. Moreover, in Italy private health expenditures increased faster than public spending from 702 $PPP in 2008 to $788 in 2018 at about one quarter of total spending. Private expenditures were mainly out-of-pocket payments, as both individual, voluntary health insurance and mutual funds in Italy remain limited in the number of organizations and of people insured. Co-payments for covered drugs and services accounted for over 90 per cent of private expenditure and, during the crisis, increased from 2.2 billion euros in 2010 to about three billion in 2018, or around 2 per cent of total health expenditures.

The imposition of a charge of ten euros for medical prescriptions was the first, almost immediate measure taken by the government in the healthcare field. This was appropriately dubbed "super-ticket" for its amount but also because it was superimposed over charges on drugs, tests, and medical visits that were now under regional regulation. This made the out-of-pocket cost of accessing SSN services a significant expenditure, particularly in a context of reduced household income and increased unemployment. The potential effect was of either postponing or skipping a visit and/or the prescription or diverting users towards private providers, when the market price of the service was similar and the inconvenience cost of accessing public services was high. There is some evidence about both effects. Data from the Survey of Health, Aging and Retirement in Europe on the citizens of eleven European countries, including Italy, show increases in both the proportion of people who incurred out-of-pocket expenditures in accessing covered services (from 64 to 80 per cent) and in mean expenditures, from 358 to 480 euros between 2006–2007 and 2013 (Palladino et al. 2016). Italy, as in Spain and the Czech Republic, had a significant increase in the proportion of people who incurred catastrophic expenses, defined as greater than 30 per cent of household income. The authors noted that the poorest segments of the population were less likely to incur catastrophic expenses than those in the highest quintile and concluded that measures could have possibly been in place for some financial protection. However, these results may also suggest that the poorest part of the population just gave up using expensive health services.

In Italy, the documented 8.5 per cent drop in the use of public specialist outpatient services could possibly be explained by shifting to private providers, postponing their use, or completely avoiding these services (Agenas 2013). The National Institute of Statistics estimated that in 2017 four million Italians, or 6.8 of the population, stopped using medical services for financial reasons (Istat 2018). This is much less than the estimated twelve million people claimed in a previous study (RBM-Censis 2017). The latter study is of some importance because it used the argument of the financial barriers imposed by the economic downturn to support the institution of a second and a third layer of mutual and individual health insurance in order "to save" the SSN. A re-emerging concept of "selective universalism" under the curious logical paradox of sustaining the SSN by abandoning its fundamental principles. The impact of austerity policies on inequality in access has been assessed across its several dimensions, including affordability, availability, approachability, acceptability, and appropriateness (Petmesidou et al. 2020). Some evidence emerged that self-reported unmet needs for medical examination due to cost, distance, or waiting list barriers increased between 2008 and 2015. This however started to decline shortly afterwards, and in 2017 rates were lower than those observed in 2008 (Petmesidou et al. 2020, p. 8).

"The obsessive quest to gut the hospital" described by Uwe Reinhardt in the United States (Reinhardt 1996) is also one of the paramount characteristics of Italian health policy. De-hospitalization has been the main policy of the SSN since its inception, first for ideological motives and then for efficiency reasons. Acute hospital beds per 1000 had been among the lowest in European and OECD countries, but Italy continued lowering national standards. In terms of efficiency, the low-hanging fruit had already been harvested, but the hope was that this would help promote primary and community care. This traditional policy was also part of the most urgent cost control measures. The spending review legislated with Law nr. 135 of 2012 to curb health expenditures set new national standards for acute hospital services at three public and/or contracted beds per 1000 population, with a rate of hospitalization of 160 per 1000 and a length of stay of less than seven days. Legislation focused on transferring inpatients to outpatient care for "low-value" admissions such as uncomplicated diabetes and hypertension, although the transfer of hospital services to ambulatory settings increased the breadth of co-payments.

The number of acute hospital beds per 1000 population fell from 3.9 in 2007 to 3.2 in 2017, close to the standard set by the health law of 2012 at 3.0 and much lower than that observed in other European and OECD countries. Admission rates also fell in the same period from 187 to 123 per 1000 population, and expenditure on hospitals declined steadily to account for 35.5 per cent of public health expenditure (Ufficio Parlamentare del Bilancio 2019). The spending review had left further specifications of the hospital reform to a Regolamento attuativo (operating rules), which had to be negotiated with the regions, now empowered with full sovereignty over the organization of health service. This took three more years and showed an unmistakable move towards the "focused factory" model in delivering hospital services (Skinner 1974; Herzlinger 1996). Going beyond service "regionalization", hospitals were to be organized and managed to concentrate on specific services or surgical procedures and/or specialize in specific diseases. Volume ceilings for performing breast cancer surgery were set at 150 procedures per year, while for cardiovascular procedures, ceiling for CABG surgery was set at 200 and for coronary angioplasty (PTCA) at 250 per year, for example. The aims were to reap the alleged benefits of volume and focus in terms of both clinical outcomes and technical efficiency, by spreading indirect and fixed costs over larger volumes of activity and extolling physicians' experience and standardization of tasks. Thus far, data are lacking to assess the extent of the implementation of this ambitious programme.

Legislation also addressed the long-awaited problem of a new organizational form of general practice, which in Italy is still dominated by solo practice with contracted physicians. The new model of primary care would entail multiprofessional group practices including other health professionals, along with a rather weak alternative of "functional" networking of solo practices. Moreover, the reform tried to combine general practice with emergency and out-of-hours services into primary care units providing 24-hour coverage. Unfortunately, this significant legislation was not implemented. Except than in a few regions, the unwavering opposition of contracted physicians and the stalling of the negotiations by their unions killed a radical reform that the government had unfortunately set to be implemented at zero cost.

Available evidence of the impact of the financial crisis on the health of the population is scarce. Overall mortality was largely unaffected, as the small increase in suicides (De Vogli et al. 2012) was offset by a decline in deaths for injuries, particularly traffic-related, which were already in steady decline (Egidi and Demuru 2018). Unhealthy behaviours such as alcohol and tobacco use remained stable, although increases in some groups already at risk have been noted (Costa et al. 2012). The most consistent effects were observed in mental health, including depression, suicides, and attempted suicides, and compulsive gambling. National surveys performed in 2005 and 2013 showed an increased risk of poor (self-reported) mental health by 17 per cent in males and 4 per cent in females (Odone et al. 2017). These were associated with unemployment and financial strains, such as mortgage and rent arrears and evictions. A significant impact of high unemployment rates on hospital admissions for severe mental affective disorders was observed in some economically more disadvantaged areas of the north (Wang and Fattore 2020). However, the health effects of the global financial crisis are still uncertain and its full impact on deaths, diseases, lifestyle, and health behaviours will possibly emerge only in the long term (Mladovsky et al. 2012; Stuckler et al. 2015; Karanikolos et al. 2016).

Reactions to the deep recession from the health field consisted primarily in emergency measures, such as the "super-ticket", and the acceleration of fiscal policies already in motion before the crisis. The government in charge failed to exploit its exceptional mandate to design realistic reforms to help reconfigure the health system and increase its resilience. When such reforms were tried, as with primary care, they went unimplemented or watered down as they met professional resistance and entered the arena of regional sovereignty. The regions accelerated the processes of horizontal and vertical consolidation in the organization of their health system. This is most evident in the fusion of various aziende sanitarie locali (ASLs), which often also incorporated aziende ospedaliere. Between 2008 and 2015, ASLs decreased in number from 154 to 139 and aziende ospedaliere from 95 to 75 (Bobini et al. 2019, p. 38). This definitively put to rest the remnants of the purchaser-provider split and greatly increased the regional grip over the system's governance. The ASLs' larger size of about half-million population on average further reduced municipalities' power in healthcare and hindered the integration with social care in the community.

The SSN has been showing an unexpected resilience adapting to the worsening fiscal climate without obvious signs of diminishing access and hampering delivery. The image of the bumblebee which allegedly flies against, or despite of, the laws of physics has been proposed (Taroni 2019). However, ten years of austerity politics in healthcare had led to an SSN that was "underfunded, underdoctored and overstretched" to borrow from the Royal College of Physicians' characterization of the British NHS (Royal College of Practitioners 2016). There were fewer nurses and doctors, and they were older and under-paid. The disjointed process of doing (more or less) the same with less money, higher co-payments, and barriers to access had not hampered the institutional logic of the SSN (Taroni 2015). However, the grip of austerity politics had both re-fuelled the old tensions between universalism and selectivity and prompted novel secessionist stances from the wealthiest regions of Lombardy, Veneto, and Emilia-Romagna. The strange alliance between two regions, Lombardy, and Veneto, governed by a centre-right majority for decades, and Emilia-Romagna, the traditional stronghold of the centre-left, asked the central government for a new and wider division of functions in favour of the regions. The devolution of additional functions, including several health issues, implied that the regions retain their "tax surpluses", that is, the difference between what the state takes out taxing the residents of the region and its transfers to the region (Ufficio Studi del Senato 2018). This would reduce the resources available to the centre to implement the redistributive policies aimed at helping the poorest regions in financing their basic functions, including healthcare. Because of its impact on the vertical dimension of equalization among the regions, the proposal has been dubbed "la secessione dei ricchi" (the secession of the rich) (Viesti 2019) and has obvious implications for the life of the SSN as a tax-financed system. In summary, a decade of austerity politics left the health system seriously under-resourced and with equally serious threats to its principles of individual and territorial solidarity. Both ultimately exacerbated the impact of the acute shock of the pandemic.

The Second Storm: The COVID-19 Pandemic

The COVID-19 pandemic struck Italy while it was still trying to recover from the Great Recession. The first two exogenous cases, tourists from China, were identified on January 31, 2020, and this prompted the declaration of the national state of emergency with a duration of six months. A few weeks later, on February 21, the Istituto Superiore di Sanità announced the identification of the first endogenous case, a thirty-eight-year-old man in Codogno, near Milan, in the Lombardy region. On the same day, a second patient was diagnosed in the small town of Vo' Euganeo, on the border of the Veneto region. An incessant stream of cases followed, punctuated by an increasing number of deaths. As of June 1, 2021, 4,201,394 cases have been officially recorded at a rate of 7090 per 100,000 population and with 124,395 deaths (Iss 2021a). Although deaths have been, and still are, concentrated in the elderly population, life expectancy at birth for both genders is now 82 years, 1.2 years less than in 2019, ranging from a reduction of 2.6 years in Lombardy to 0.5 years in Calabria (Istat 2021a). This is the heaviest toll of any event in Italy since the Second World War and one of the highest in the world.

Figures 11.2 and 11.3 display the general trend of the pandemic through June 1, 2021, in terms of cases and deaths, respectively. Sixteen months from its outbreak, the "national" epidemic has proved far from homogeneous both in space and in time. The curve of the first wave surged in February 2020, peaked in March, and was flattened in late May by a long period of lockdown. A low incidence phase followed, lasting until early September, when a second wave hit. The epidemic's distribution in space can be better described as patchy, with several hotspots across the nineteen regions and two autonomous provinces emerging at different times. The first epicentre of the epidemic was concentrated in a few northern spots, while some regions, particularly in the southern part of the country, had no reported cases until late March. Some still have relatively few cases, but others, such as Campania, have experienced a staggering surge of infections. In June, some of the most sparsely populated areas in the Alps (Valle d'Aosta, Trento, and Bolzano) ranked very high in cases per capita (Iss 2021a) (Table 11.1). After a stumbling start at the end of December 2020, a vaccination programme has covered most, if not all, of

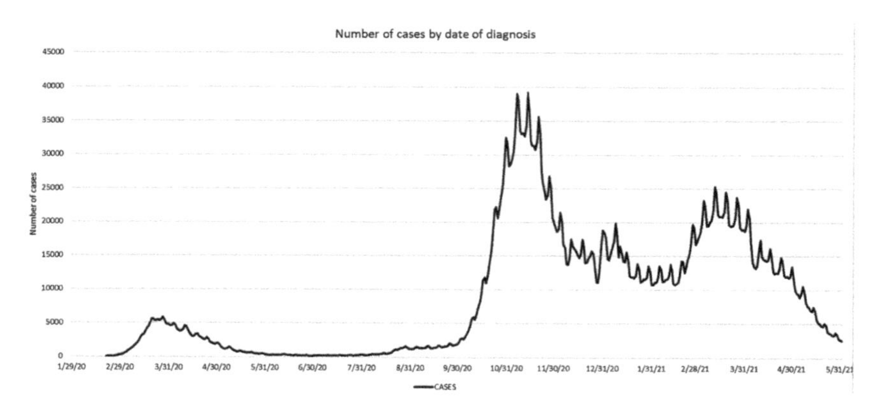

Fig. 11.2 Number of reported COVID-19 cases, February 2020–June 2021. (*Source*: Istituto Superiore di Sanità Open Data www.iss.it)

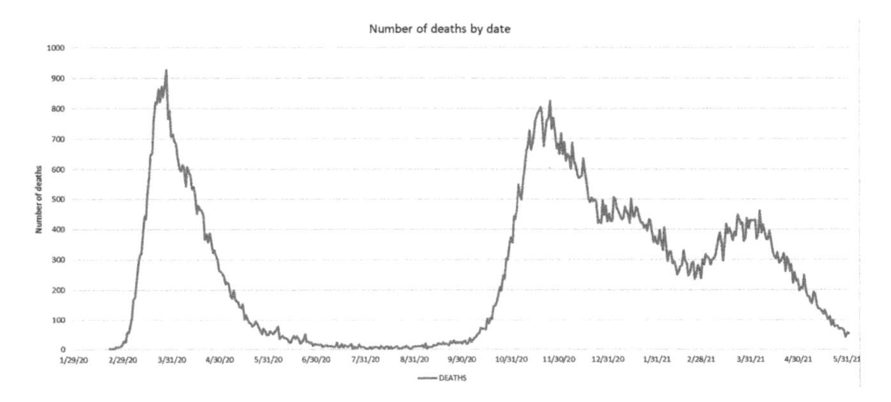

Fig. 11.3 Number of COVID-19 deaths, February 2020–June 2021. (*Source*: Istituto Superiore di Sanità Open Data www.iss.it)

Table 11.1 Ten regions with the highest rates of reported COVID-19 cases as of June 1, 2021

Region	*Cumulative incidence/100,000 residents*
P.A. Bolzano	12,066
Val d'Aosta	9,307
Veneto	8,742
Emilia-Romagna	8,635
P.A. Trento	8,534
Piedmont	8,474
Lombardy	8,372
Campania	7,090
Liguria	6,816
Marche	6,776
Italy	7,090

Source: Istituto Superiore di Sanità (2021a)

the higher-priority groups (healthcare workers, residents of care homes and their caregivers, the elderly and medically vulnerable groups) and is now addressing the youngest segments of the population. The vaccination campaign is accelerating as supply increased and its incentives refined. To help booster vaccination rates, a "Green Pass" has been introduced at work and schools, to enter restaurants and theatres, and, starting in the fall, to use long-distance trains and airplanes. This is contributing to a much-anticipated fall in the number of cases, hospitalizations, and deaths.

However, as of this writing, cases are spiking again across all Italian regions in a surge that exceeds last year's summer wave in the number of cases. They tend to be younger and healthier than before, and case-fatality and death rates are lower.

Sixteen months into the pandemic, our poor understanding of the disease and its capricious evolution make forecasts hazardous. It makes it difficult to interpret even the recent past. Reported cases are unreliable as measures of the spread of the infections and as indicators of the response of the regional systems. Serological investigations suggest that case numbers are a fraction of the total number of people infected, currently estimated at one reported case for every six infected people (ISTAT-Iss 2020). Reported cases also depend on testing capacity, which was slow to begin in Italy, and varied greatly among the regions and over time. This is not unprecedented. For example, as with the 1918 flu, whose neat separation in different waves is the product of the posterity's hindsight (see Chap. 3), it is not clear if Italy experienced two or three different waves of the pandemic of COVID-19. It therefore seems useful to consider the development of the COVID-19 pandemic in Italy in three phases and two distinct waves, whose main characteristics are summarized in Table 11.2.

Writing "the history of the present" (Berridge 1994) over a very politically and emotionally sensitive subject risks a kind of "slow journalism"

Table 11.2 COVID-19 pandemic: tests, new cases, and deaths, first and second waves (average rates per week per 100,000 population)

	First wave[b]	*Second wave*[c]
Number of Tests[a]	380.0	2473.7
New cases[a]	36.7	192.3
• North	64.2	230.1
• Centre	21.4	174.7
• South	5.8	156.7
Deaths[a]	5.0	4.4
• North	9.5	5.5
• Centre	2.1	3.9
• South	0.8	3.2

[a]per 100,000 population per week

[b]2020/2/24–2020/5/4

[c]2020/9/15–2021/4/25

Source: Istituto Superiore di Sanità Open data www.iss.it

(a definition Virginia Berridge attributes to the American political scientist and historian Daniel Fox writing on the history of AIDS). To maintain a little distance from the events, we can only consider what we legitimately would call the first wave of the pandemic, spanning from the first cases and the declaration of the state of emergency to mid-May, when restrictions were eventually lifted. The following pages briefly summarize the cascade of events, which led to the COVID-19 tragedy in terms of deaths and economic and social disarray. We then review the dynamics of the policy response to the epidemic, including the interplay between scientific and technical expertise and policy choices as well as the strain that the asymmetrical spread of the epidemic imposed on the relations between the central government and the regions.

The pandemic started in the provinces of Lombardy and their neighbours with established trade connections with China and then spread rapidly through these areas with high population density and frequent internal travel and trade. Mass gathering events also helped, as with the famous Champion football game that an estimated 50,000 people from Bergamo attended. By February 24, out of the 229 cases reported to the Ministry of Health 213 were from three regions: Lombardy reported 172 cases, Veneto 33, and Emilia-Romagna 18. These three regions accounted for about three-quarter of cases and over 70 per cent of excess mortality of the first wave and still account for 46, 13, and 9 per cent respectively of all Italian cases (Iss 2021a).

The surge was dramatic in some provinces of the Lombardy region, including Bergamo, Brescia, and Cremona, which at the time had one-third of the total cases in the country. The unremitting number of infected people who presented themselves simultaneously rapidly overburdened, overstretched, and eventually overwhelmed the richly resourced health systems of the three regions, both in their community capacity of containing the spread of the infection and in providing hospital care. Community services stopped contact tracing claiming a lack of resources and testing capacity. Hospitals reorganized to cope with the influx of severe patients. A vivid description of the grim situation in the hospital of Bergamo, a large teaching hospital at the epicentre of the first crisis, is reported by Nacoti and other hospital doctors, including a group of Cuban colleagues (Nacoti et al. 2020), and thoughtful comments on their difficult clinical choices are in Rosenbaum (Rosenbaum 2020). The collapsing hospital capacity, and particularly the impending or actual saturation of Intensive Care Units (see, e.g., Grasselli for the Lombardy region—Grasselli et al. 2020), was the critical factor inducing the central government to adopt strict stay-at-home orders to mitigate the spread of the infection. On March 9, Italy was the first European country to adopt strict containment measures including closure of schools and of all non-essential businesses, suspension of public events and other gatherings, and prohibition of circulation except for essential work activities. The early positive effects of the lockdown showed up in mid-April when the incidence of new cases and deaths started to steadily decline. The lock-down was relaxed and then lifted by mid-May, as with most of other European countries. Bosa et al. (2021) provide a timeline of national policy responses to the first wave of the pandemic.

Figures 11.2 and 11.3 show a rather long interval corresponding to the summer of 2020 that neatly separates the spring wave of 2020 from the second wave. A few scattered foci of contagion appeared throughout the summer, but by late August an epidemic resurgence was clear. A distinct second wave started in September, worsening as fall headed into winter. The dramatic acceleration in the number of cases and deaths was both sudden and sustained. The fall in case-fatality rates from 6.6 per cent in the first wave to 2.4 per cent reported by the *Istituto Superiore di Sanità* through October (Iss 2021b) is due, in part, to the increase in testing capacity and its broader application to suspect any asymptomatic cases. In this wave, infections were initially concentrated in younger people, who suffered a milder form of the disease, but then spread the infection to their parents and grandparents. Along with the substantial increase in testing capacity, this explains, in part, the lower death rates. Unlike the first wave that mostly struck the country's more affluent north, the second surge also reached the southern regions, with weaker economies and less rich health systems and hospital networks. Moreover, many of them had been subject to *piani di rientro* (see above) and had suffered cuts in the number of hospital beds and personnel freezes.

Beginning in mid-September, no abrupt reduction in the number of cases is observable, except a short drop in the number of cases correspond-ing to the turn of the year, associated with a much shorter period of lock-down during Christmas festivities. The steadier number of cases observed during this period is again possibly associated with the massive increase in the number of tests performed, which now included antigen testing. Beginning with this second wave, Italy avoided long and generalized lock-downs, implementing more targeted, regionally based restrictions through a three-tiered system of red, orange, and yellow zones. A fourth colour, the white zone, was for regions declared "COVID-free". The new approach, like that adopted in England (Hunter 2020), was intended to be a proportionate reaction to varying severity of the epidemics across the regions to best preserve economic activities where the pressure on hospitals and particularly ICU beds was lower.

In the first wave of the pandemic, Italy experienced higher COVID-19 associated deaths and excess all-cause mortality than those observed in other countries. Italy's cumulative per capita COVID deaths at 59.1 per 100,000 population observed in September 2020 since the start of the epidemics compared poorly with France (46.6) and put Italy among the countries with the highest death rates, including the United States (60.3), Britain (62.2), and Spain (65.5) (Bilinski and Emanuel 2020). Based on data through mid-March 2020, the case-fatality rate of persons with confirmed COVID-19 was 7.2 per cent (Onder et al. 2020). Out of 27,967 deaths officially reported in the first phase, the mean age was seventy-nine years. Explanations include the fact that Italy has one of the oldest populations in the world, second only to Japan, and this was particularly the case in the regions where most cases concentrated in the early phase. The low frequency of testing and missed diagnosis of the disease, particularly in patients who died in care homes or at home, could also have negatively biased the estimated case-fatality rate compared to other countries. Excess mortality, that is, deaths beyond what would be expected during the pandemic, has consistently been far higher than the number of deaths officially reported as associated with COVID-19 infection. The National Institute of Statistics found 45 per cent excess mortality during the first wave, 60 per cent of which was associated with COVID (Istat 2021b). Between February 15 and May 15, 2020, 47,490 deaths (29. 5 per cent of total deaths in the period) in excess from the expected baseline were observed, compared to 31,610 official COVID-associated deaths (Scortichini et al. 2020). Excess mortality showed a strong geographical pattern, much higher in the northern regions, where Lombardy, Veneto, and Emilia-Romagna accounted for 71 per cent of excess deaths, milder in central regions and smaller or absent in the south.

Residents of care homes, variously defined, were a significant proportion of total official COVID deaths. A national survey of care facilities for the elderly and the disabled found that 41.5 per cent of deaths in these facilities were associated with COVID (Iss 2020). Despite an early understanding that the frail elderly with comorbidities were at higher risk of death, residential facilities were not prioritized in the early phase of the pandemic, as it also happened in several other developed countries (Comas-Herrera et al. 2020), includind the United States (Grabosky 2020) and Canada (Marchildon and Tuohy 2021). The virus spread from the community via visitors and staff and diffused widely in the absence of proper testing, appropriate procedures, and isolation of residents and staff. The transfer of patients from acute hospitals also significantly contributed. Although residential facilities have suffered from the general shortage of testing facilities and personal protective equipment (PPE), in Italy as elsewhere, the crisis has also exposed long-standing shortcomings in how they are structured, staffed, and financed. Confined spaces and physical layouts that inhibit isolation; staff working at low pay and with little training in dealing with infectious diseases; poor or absent information systems and extremely weak regulatory oversight transformed residential facilities for the elderly and vulnerable populations into hotspots of the COVID pandemic. In Canada, the clustering of COVID-19 deaths in long-term care facilities spurred proposals for including long-term care in Medicare's universal health coverage and imposing stricter accreditation measures upon facilities and staffing competences (Marchildon and Tuohy 2021). Such sensible proposals seem still lacking in Italy.

In the first wave of the pandemic, a dramatic surge in the number of cases swept through the limited resources of community services for containing the spread of the epidemic through testing and tracing and rapidly overflowed hospitals. The standard capacity of the hospital system in terms of Intensive Care Unit (ICU) beds at 12.5 per 100,000 population was in line with other European countries (11.5), although much lower than in Germany and the United States, at 29.2 and 28.0 ICU beds per 100,000 population (Rhodes et al. 2012). Repeated slashing of acute hospital beds had not significantly reduced their availability (Pecoraro et al. 2020; see also Bosa et al. 2021). Hospital facilities reorganized, including the creation of specialized COVID units, and some COVID hospitals, similar to the Nightingale Hospitals in England, were created. Reorganization included the transformation of operating rooms into ICUs, which required halting elective and non-emergency surgery and led to a backlog of cases. Existing staff were redeployed, regardless of their specific competencies, and recently retired doctors and nurses were brought in, as well as newly graduated medical staff. Disrupted processes, fatigue, and the shortage of personal protective equipment helped spread the infection within the hospitals and exposed staff to high risk. Routine activity levels were disrupted along the entire pathway of early diagnosis, initial treatment, and follow-up for cancer, cardiac diseases, transplants, and other serious conditions.

When the pandemic struck, the SSN was still trying to recover from the period of zero or negative growth during the Great Recession. As we have seen, cutbacks had focused on the workforce, which had diminished in number and pay, and was older. This attrition had stripped the health system of much of its resilience, particularly in the crucial sector of community services that had the fundamental role of containing the epidemic through contact tracing and of home care for the milder cases. As emergency measures, the government increased the resources available for hospitals and community services, enrolling community nurses and instituting home care teams. However, this was too little and, above all, too late, and hospitals were overwhelmed. Community services, the first line of the response to epidemics, essentially failed. They failed in various ways across the regions, which brings us to the issue of the relations between the state and the regions during the pandemic. The management of the pandemic was essentially a stress test of the relations between central and regional governments over the alleged model of cooperative regionalism, and this revealed serious problems of coordination around two main problems. The first was how to combine the local implementation of national uniform measures with a practice of strong regional autonomy in healthcare organization and varying degrees of performative capacity. Additional coordination problems were exacerbated by the varying and frequently changing rates of infection across the regions. Moreover, the regions most affected by the pandemic in the first wave are also the richest, in that they "own" 40 per cent of Italian GDP and exert a large political influence over the central government. It also happens that they are the regions that had requested the special procedure for differentiated regionalism.

Both Constitutional and health legislation clearly state the supremacy of the national government over the control of epidemics. A recent pronouncement of the Constitutional Court states it quite firmly (Cuocolo and Gallarati 2021). The national government declared the state of emergency on January 31, 2020, with a duration of six months, as required by the Constitution. The Department of Civil Protection (DCP), under the direct authority of the Prime Minister, was given operational authority, as commonly happens in case of disasters. However, since the DCP has no expertise or experience in epidemics, a panel of scientists including epidemiologists, virologists, and the like provided advice. Furthermore, a commissioner with special powers was appointed to coordinate the procurement of PPE and ventilators and later of vaccines, and to organize the mass vaccination campaign. In short, the management of the pandemic was framed as a highly technical problem that required quick and expert-led decision-making at the highest political level. In theory, this set up a very centralized scheme of governance potentially capable of swift decisions and prompt action, but things transpired differently.

The high degree of regional autonomy in the organization and management of health services entered into competition with national powers. Calling upon their veto power as the sole actors of their implementation, the regions negotiated national measures to respond to the epidemic. The adoption of the three tiers of restrictions, for example, was one of the results of such negotiations, which led to further politicization of the COVID response along regional lines. As the central government restrained itself in issuing legislation and focused principally on broad mitigation measures and stay-at-home orders, each region developed its own response to its epidemic, with different strategies.

Initially, the government's declaration of the state of emergency appeared to be out of proportion with respect to the present danger. This reaction is all too frequent in the early phases of the epidemics and explains why it is difficult to make decisions when the crisis is still unfolding (Pisano et al. 2020). This led to important philosophical debates over the "state of exception" and its impact on personal freedoms and civil rights (Agamben 2020). More importantly, in terms of its implementation, the declaration of the state of emergency was met by scepticism by the major political and economic interests, and poorly thought-out initiatives such as "Bergamo is running" and "Milan won't stop" were taken to counter the measures of mitigation adopted by the central government (Imarisio et al. 2020, p. 78 and 80 respectively). The management of the crisis turned into a round of blame games, where the opposition first pushed for reopening the "red zones" in the Lombardy and then mobilized to close the whole country (Capano 2020). Local and national politicians blamed the conflicting opinions of the "scientists" over the course of the pandemic; some openly opposed masking and supported unconventional treatments such as hydroxychloroquine as an alternative to lockdowns.

Relaxation of the pandemic restrictions marked a change in the public debate, which focused on the economic impact of the pandemic and the financial support of the economic sectors and categories most affected. The regions voiced local discontent and claimed extensive powers in supporting local economies. The influence of political parties also emerged when redistributive issues were at stake. The result of the clash of these countervailing forces was an (apparently) incremental strategy riddled with inconsistencies, which tried to balance health protection of the population with a much too large spectrum of economic and political interests.

The clash of political and economic interests is an important dimension of the relations between the central government and the regions over the management of the pandemic. This is the focus of early analyses of regional policies such as Capano (2020) and Capano and Lippi (2021). The regional management of the pandemic also brought into sharp focus the differences in the structure and functioning of their health systems. The most notable example is the contrast between the management strategies of Lombardy and Veneto, two neighbouring regions, both governed for decades by a centre-right coalition led by the *Lega Nord*. Veneto took proactive measures aimed at containing the spread of the virus, with an aggressive strategy of contact tracing and testing and home care (Binkin et al. 2020). Compared to Lombardy, this resulted in twice as many tests per 1000 population (23 vs. 12), fewer hospitalizations (25 vs. 51 per cent), and lower mortality rates (10 vs. 75 per 100,000). The strategy benefitted from the priority given to the surge in testing capacity but was based on the strong public health infrastructure of the Veneto region. Lombardy had eight public health departments, or one for every 1.2 million population, compared with one per 500,000 population in Veneto (Binkin et al. 2020). Several other factors may explain the differences between the two regions in the early phase of the pandemic. These include, for example, the higher population density in Lombardy and the, yet unproven, hypothesis that the virus was circulating there well before the first case was detected. Participation of the general practitioners in contact tracing and home care is also an important issue, more debated than studied. Furthermore, the potential for differential stages in the development of the epidemic, as well as for system performance to change over time must be factored in any assessment. Measures to slow transmission and reduce mortality depend on government measures and actions, public health capacity, and community behaviours (Fisher et al. 2020). All these can change over time, along with the pandemic. Certainly, something happened between then and now in one or more of these dimensions to explain why Veneto now has the same cumulative incidence of cases as Lombardy. It seems plausible, however, to argue that the apparently more proactive, community-oriented policy performed better than a more passive, hospital-centred approach in terms of cases, hospitalizations, and deaths. The traditional public health strategy of case finding, tracing, and testing seems essential in integrating physical distancing and other so-called non-pharmaceutical interventions to curb the spread of the infection (Saracci 2020).

When the pandemic hit Italy, the IMF had predicted a return of the GDP to pre-crisis levels by mid-2020 and the SSN had begun to see a slight increase in its budget after a decade of zero or negative growth (Fig. 11.1). Instead, in the year 2020, GDP declined by 8.9 per cent, and the public debt rose from 134.8 per cent of GDP in 2019 to a staggering 158 per cent (Pianta et al. 2021). Because of the closing of non-essential businesses, an estimated eight million workers became temporary unemployed. To mitigate the economic and social impact of the pandemic, including the artificial coma of the economy induced by stay-at-home orders, the government has so far distributed about eighty billion euros in subsidies and incentives. SSN expenditures have also greatly increased, with the extra-funding allocated to consumables and vaccines, as well as to the temporary hiring of new personnel. However, a substantial share of this extra-funding is non-recurrent, and projected "normal" budgets for the next few years show the usual decline.

The European Union has also changed its austerity policy, suspending the Stability and Growth Pact and passing the historic Next Generation EU to help repair the economic and social damages of the pandemic. The Plan, which entails elements of debt mutualization, is a new departure for the EU, born out of the new cooperative policy of the European Union. Italy is the single largest beneficiary of the Plan, with 196.5 billion euros. Yet another technocratic government, led by Mario Draghi, former President of the European Central Bank, is responsible for managing the Plan (Bull 2021; Garzia Karremans 2021). The pivotal strategy for the recovery of the SSN is integral to the national recovery and is broadly described in a chapter of the "Recovery and Resilience Plan: Next Generation Italia" (www.mef.gov.it). Of the total 223.9 billion euros allocated to the six missions of the plan, 19.72 billion euro will finance healthcare. A major component of the Health Mission is devoted to "Proximity Assistance and Telemedicine" aimed at strengthening community and home care. This includes bringing integrated social and health services closer to where people live by developing community networks and modernizing technologies. This is one of the main shortcomings of the current organization of the SSN, which the pandemic made apparent. As we have seen, this was also the principal issue of the debate at its dawn and the best part of the tradition of social medicine in Italy.

Social and economic recovery will certainly be extremely demanding: deficit spending has reached unprecedented levels and GDP has fallen dramatically. The effectiveness of the Plan in repairing the poor status of the SSN, however, will also depend on the characteristics and future developments of the pandemic. Now well into its second year, the early framing of the COVID-19 pandemic as a sudden, time-limited visitation of a virulent external agent, which has been developed in analogy with the devastating pandemics of the past, is questionable. The unremitting series of incoming waves, the emergence of variants, and the still poorly understood "long COVID" suggest developing long-term strategies for responding to a long-standing condition rather than adopting emergency measures to fight the time-limited visitations of an infectious disease. As happened with the history of AIDS masterfully examined by Nancy Krieger and Elizabeth Fee (Fee and Krieger 1993a, b), we may expect different phases in the unfolding of the COVID-19 pandemic, each with very different implications for health and social policy, including patterns of services, financing, and research.

References

Agamben, G. (2020) A che punto siamo? L'epidemia come politica. Macerata, Quod libet

AGENAS (2013) Gli effetti della crisi economica e del superticket sull'assistenza specialistica. Roma, AGENAS

Bobini, M., Cinelli, G., Gregiatti, A., Petracca, F. (2019) La struttura e l'attività del SSN. In: Rapporto OASI 2019. Milano, Egea, pp. 33–68

Berridge, V. (1994) Researching contemporary history: AIDS. History Workshop, 38, pp. 228–234

Binkin, N, Michieletto, F., Salmaso, S., Russo, F. (2020) Protecting our health care workers while protecting our communities during COVID-19 pandemic: A comparison of approches and early outcomes in two Italian regions. Med. RXiv. https://doi.org/10.1101/2020.04.10.20060707

Bilinski, A. Emanuel E.J. (2020) COVID-19 and excess all-cause mortality in the uS and in 18 comparison countries. JAMA Nov.24;324(20), pp. 2100–2102

Bosa, I. Castelli, A., Castelli, M. et al. (2021) Response to Covid-19: was Italy (un)prepared ? Health Economics, Policy and Law. https://doi.org/10.1017/s1744133121000141

Bull, M.J. (2021) The Italian government response to Covid-19 and the making of a prime minister. Contemporary Italian Politics, 13, pp. 149–165

Bull, M.J., Pasquino, G. (2018) Italian politics in an era of recession: The end of bipolarism? South European Society and Politics, 23 (1), pp. 1–12
Capano, G. (2020) Policy design and state capacity in the COVID-19 emergency in Italy: if you are not prepared for the (un)expected, you can be only what you really are. Policy and Society, 39 (3), pp. 326–344
Capano, G. Lippi, A. (2021) Decentralization, policy capacities, and varieties of first health response to the Covid-19 outbreak: evidence from three regions in Italy. Journal of European Public Policy. 28, pp. 1197–1218
Costa, G., Marra, M., Salmaso, S., Gruppo AIE su crisi e salute (2012) Gli indicatori di salute ai tempi della crisi in Italia. Epidemiologia & Prevenzione, 36 (6), pp. 337–366
Comas-Herrera, A., Zalakonn, J., Lemmon, E. et al. (2020) Mortality associated with Covid-19 outbreaks in care homes: Early international evidence. LTC covid.org. International Long-Term Care Policy Network, CPEC-LSE
Cuocolo, L., Gallarati, F. (2021) La Corte difende la gestione unitaria della pandemia con il bazooka della profilassi internazionale. Corti Supreme e Salute, 1, pp. 1–18
De Vogli, R., Marmot, M., Stuckler, D. (2012) Excess suicides and attempted suicides in Italy attributable to the great recession. Journal of Epidemiology and Community Health, 67 (4), pp. 378–379
Economou, C., Kaitelidou, D., Kentikelenis, A.E., Sissouras, A., Maresso, A. (2014) The impact of the financial crisis on health and the health system in Greece. In: Maresso A., et al., Eds. Economic Crisis, Health system and health in Europe. Country experiences. Copenhagen, WHO, pp. 103–142
Egidi, V., Demuru, E. (2018) Impatto delle grandi crisi italiane su salute e mortalità: Il caso italiano. In: La società italiana e le grandi crisi economiche, 1929–2016. Annali di Statistica. ISTAT, pp. 57–82
Emmott, B. (2012) Good Italy, Bad Italy. New Haven, Yale University Press
Fee, E., Krieger, N. (1993a) The emerging histories of AIDS: three successive paradigms. History of Philosophy and the Life Sciences, 15, pp. 459–487
Fee, E., Krieger, N. (1993b) Thinking and rethinking AIDS: implications for health policy. International Journal of Health Policy, 23, pp. 323–346
Fisher, D., Teo, Y.Y., Nabarro, D. (2020) Assessing national performance in response to COVID-19. Lancet 396, pp. 653–655
Garzia, D., Karremans, J. (2021) Super Mario 2: comparing the technocrat-led Monti and Draghi governments in Italy. Contemporary Italian Politics, 13, pp. 105–115
Grabowski, D.C. (2020) Nursing home case in Crisis in the wake of COVID-19. Journal of the American Medical Association 324, 23–24.
Grasselli, G., Pesenti, A., Cecconi, M. (2020) Critical care utilization for the COVID-19 outbreak in Lombardy, Italy: Early experience and forecast during an emergency response. Journal of American Medical Association, 323 (16), pp. 1545–1546

Greer, S.L. (2014) The three faces of European Union health policy: Policy, markets and austerity. Policy and Society, 33, pp. 13–24

Herzlinger, R. (1996) Market-driven healthcare: Who wins, who loses in the transformation of America's largest service industry. Reading, MA, Addison-Wesley

Hunter, D.J. (2020) Trying to "Protect the NHS" in the United Kingdom. New England Journal of Medicine, 383, e136

Imarisio, M., Ravizza, S., Sarzanini, F. (2020) Come nasce un'epidemia: La strage di Bergamo, il focolaio più micidiale d'Europa. Milano, Rizzoli

Iss (2020) Survey nazionale nel contagio COVID-19 nelle strutture residenziali e sociosanitarie. Report finale. Aggiornamento 5 Maggio.Roma (www.iss.it)

Iss (2021a) Epidemia COVID-19.Aggiornamento nazionale. 1 Giugno 2021. (www.iss.it)

Iss (2021b) Characteristics of SARS-COV-2 patients dying in Italy. Report based on available data on April 28, 2021. Roma (www.iss.it)

Istat (2018) Indagine sugli aspetti della vita quotidiana, anno 2017. Roma

ISTAT-Iss (2020) Impact of the COVID-19 pandemic on the total mortality of the resident population in the first quarter of 2020. https://www.istat.it/it/files/2020/05/Istat-iss_eng.pdf

ISTAT (2021a) Indicatori demografici. Anno 2020. Report 3 Maggio 2021.Roma (www.istat.it)

ISTAT (2021b) Prima ondata della pandemia. Un'analisi della mortalità per causa e luogo del decesso. Marzo-Aprile 2020. Report 21 aprile 2021. Roma (www.istat.it)

Karanikolos, M., Heino, P., Mckee, M., Stuckler, D., Legido-Auigley, H. (2016) Effects of the Global Financial Crisis on health in high-income OECD countries: A narrative review. International Journal of Health Services, 46 (2), pp. 208–240

Marangoni, F. (2012) Technocrats in government: The composition and legislative initiatives of the Monti government eight months into its terms of office. Bulletin of Italian Politics, 4 (1), pp. 135–149

Marchildon, G.P., Tuohy, C.H. (2021) Expanding health coverage in Canada: a dramatic shift in the debate. Health Economics, Policy and Law, 16, pp. 371–377

MEF—Ministero dell'Economia e della Finanza (2016) Il Monitoraggio della Spesa sanitaria, Rapporto n° 3, Roma

Mladovsky, P.D., Srivastava, J., Cylus, M., Karanikolos, M., Everovits, T., Thomson, S., McKee, M. (2012) Health policy responses to the financial crisis in Europe: Policy Summary Copenhagen, WHO/European observatory on Health System and Policies

Morgan, D.R., Astolfi, R. (2015) The financial impact of the GFC: health care spending across the OECD. Health Economics, Policy and Law, 10 (1), pp. 7–19
Nacoti, M., Ciocca, A., Giupponi, A., et al. (2020) At the epicenter of the COVID-19 pandemic and humanitarian crisis in Italy: Changing perspectives on preparation and mitigation. NEJM Catalyst, 1 (2),pp. 1-5
Odone, A., Londriscina, T., Arnerio, A., Costa, G. (2017) The impact of current economic crisis on mental health in Italy: evidence from two representative national surveys. European Journal of Public Health, 28 (3), pp. 490–495
Onder, G., Rezza, G., Brusaferro, S. (2020) Case-fatality rate and characteristics of patients dying in relation to COVID-19 in Italy. JAMA May 12; 323 (18), pp. 1775–1776
Palladino, R., Lee, J.T., Filippidis, T., Millett, C. (2016) The Great Recession and increased cost sharing in European health systems. Health Affairs, 35 (7), pp. 1204–1213
Pavolini, E., Leon, M., Guillen, A.M, Ascoli, U. (2015) From austerity to permanent strain? The EU and welfare state reform in Italy and Spain. Comparative European Politics, 13, pp. 56–76
Pecoraro, F., Clemente, F., Luzi, D. (2020) The efficiency in the ordinary hospital bed management in Italy: An in-depth analysis of intensive care units in the areas affected by COVID-19 before the outbreak. PLOS ONE, 15 (9) e 0239249
Petmesidou, M., Guillen, A.M., Pavolini, E. (2020) Health care in post-crisis South Europe: Inequalities in access and reform trajectories. Social Policy & Administration 54 1–18
Pianta, M., Lucchese, M., Nascia, L. (2021) The Italian government's economic-policy response to the Coronavirus crisis. Contemporary Italian Politics, 13, pp. 210–225
Pisano, G.P., Sadun, R., Zanini, M. (2020) Lesson from Italy's response to coronavirus. Harvard Business Review, March 27
Quaglia, L. (2009) The response to the Global Financial Turmoil in Italy: "A financial system that does not speak English". South European Society and Politics, 14 (1), pp. 7–18
RBM—Cesnsis (2017) VII Rapporto RBM- Censis sulla Sanità pubblica, privata e intermediata: Il futuro del Sistema Sanitario in Italia tra universalismo, nuovi bisogni di cura e sostenibiltà. Roma
Reinhardt, U.E. (1996) Spending more through "Cost Control": Our obsessive quest to gut the hospital. Health Affairs, 15 (2), pp. 144–145
Rhodes, A., Ferdinande, P., Flaatten, H., et al. (2012) The variability of critical care bed numbers in Europe. Intensive Care Medicine, 38, pp. 1647–1653
Rosenbaum, L. (2020) Facing COVID-19 in Italy. Ethics, logistics and therapeutics on the epidemic's front line. New England Journal of Medicine, 382 (20), pp. 1873–1875

Royal College of Physicians (2016) Underfunded, underdoctored, overstretched: the NHS in 2016. London, Royal College of Physicians

Saracci, R. (2020) Learning from COVID-19: Prevention is a strategic principle, not an option. American Journal of Public Health, 110 (12), pp. 1803–1804

Scortichini, M., Schneider das Santos, R., De Donato, F., et al. (2020) Excess mortality during the COVID-19 outbreak in Italy: A two-stage interrupted time-series analysis. International Journal of Epidemiology, 49, pp. 1–9

Skinner, W (1974) The focused factory. Harvard Business Review, May-June, pp. 113–122

Stuckler, D., Reeves, A., Karanikolos, M., Mckee, M. (2015) The health effects of the global financial crisis: can we reconcile the different views? A network analysis of literature across disciplines. Health Economics, Policy and Law, 10 (1), pp. 83–89

Taroni, F. (2015) Health care policy and politics in Italy in hard times. Journal of Health Services Research and Policy, 20 (4), pp. 199–200

Taroni, F. (2019) Il volo del calabrone: 40 anni di Servizio sanitario nazionale. Roma, Il Pensiero Scientifico

Ufficio Parlamentare di Bilancio (2019) Lo stato della sanità in Italia. Focus Tematico n° 6.

Ufficio Studi del Senato (2018) Il regionalismo differenziato e gli accordi preliminari con le regioni Emilia-Romagna, Lombardia e Veneto. Dossier n° 16, Roma.

Vicarelli, G., Pavolini, E. (2015) Health workforce governance in Italy. Health Policy, 119 (12), pp. 1606–1612

Viesti, G. (2019) Verso la secessione dei ricchi? Autonomie regionali e unità nazionale. Bari-Roma, Laterza

Wang, Y, Fattore, G. (2020) The impact of the great economic crisis on mental health care in Italy. European Journal of Health Economics, 21, pp. 1259–1272

CHAPTER 12

Postscript

Health and healthcare policy and politics are among the most distinctive characteristics of a country. They are at the forefront of public policies, part of the political, institutional, cultural, and social history, causes and consequences of economic growth and modernization of the state and civil society. This book highlights the transformations of health policy in the 160 years from the unification of Italy through the ongoing (as of this writing) COVID-19 pandemic. It describes the scale and pace of change, as well as the continuity across the three regimes (liberal, fascist, and republican-democratic) that characterize Italy's political history. This places the history of health, healthcare, and their institutions firmly in the political and social context in which they developed and evolved. Comparisons with other countries (mostly Britain, but also the United States and France) do not aim at ranking performances or draw generalizations of theoretical import. The book adopts a comparative approach in the sense that uses comparative information from various countries and the different Italian regimes to examine the strategies adopted by epistemic communities and decision makers to deal with similar problems in different contexts. Contrasting health policies over a broad arc of time and under different political regimes and comparing Italy with other countries provide some groundwork for further studies to help understand the dynamics of policy change and its impact over population health.

F. Taroni, *Health and Healthcare Policy in Italy since 1861*,
https://doi.org/10.1007/978-3-030-88731-5_12

Regime Types and Health Policies and Institutions

Liberal Italy had principally a municipal system of healthcare, which was legislated by the central government but administered and funded by *comuni*. It is telling that Giolitti's amendments to Crispi's public health law referred to "the inhabitants of the *Comuni* of the Kingdom", and not collectively to national citizens or the members of the Kingdom, also called, at the time, "*regnicoli*". Healthcare organization envisaged a residual system for the local poor, where provision of domiciliary care was provided by medici condotti under contract with municipalities, while a maze of private, mostly religious, charities or Opere pie provided residential care. This apparently decentralized scheme was substantially different from the British system of local rates in a context of national "permissive legislation". In Italy, power remained heavily centralized, and local financial resources were primarily based on regressive excise taxes, including the most hated grist tax, or *tassa sul macinato*. A further difference was that in Italy mutual aid societies were small in size and few in number, organizing predominantly the wage earners and artisans of the large, wealthy cities of the northern part of the country. They excluded significant segments of the working population. Women were usually not admitted, and the poorest workers were not eligible because of their fluctuating and insufficient income.

This approach offloaded the burden of expenditures for sanitary works and poor relief from the state to municipalities. Charities were a de facto part of the system, as health was an obligatory municipal expenditure only when "alternative" sources of funding were not available. Despite the ideological aim of their secularization by the fiercely anticlerical state, financial considerations provided compelling reasons for watering down effective municipal intervention in the activities of charities. Most local administrations were endowed with neither the financial resources nor the administrative and technical capacity to effectively manage sanitary works and poor relief services. This resulted in slow development and increasing disparity between municipalities in the northern and southern parts of the country, because of the poverty trap in which poorer municipalities had fewer resources to extract and fewer charities to rely upon. For education, the Daneo-Credaro law of 1911 gradually transferred cost and responsibility from local government to the state. However, this never happened for health, increasing the significant health inequalities between the "*due Italie*" (two Italies) of the north and the south of the country.

For about half a century, encompassing the twenty years of fascist dictatorship and the first thirty years of the democratic republic born after the Second World War, Italy adopted a mutualist system of Bismarckian ancestry.

The fascist corporatist, occupationally based sickness insurance schemes slowly emerged along the path set by Carta del Lavoro, approved on April 21, 1927. Insurance against tuberculosis was made compulsory for occupational categories at risk in 1927, and by 1929, over eight million workers and their relatives were enrolled. The insurance was centrally administered by the *Istituto Nazionale Fascista della Previdenza Sociale* (INFPS), the financial powerhouse of the regime instituted in 1933, following the scheme of "parallel administrations" invented during the liberal period. INFPS autonomously developed a separate and comprehensive system of residential care, building a very large and much celebrated network of sanatoria that by the end of 1939 encompassed more than 16,000 beds. Insurance against malattie comuni was much slower to develop. Negotiated by the organizations of employers and employees under their corporatist arrangements, separate schemes for various occupational groups were governed by a maze of several hundred semi-autonomous mutual funds. The spread of casse mutue enshrined in sectorial labour contracts privileged the industrial workers of the northern regions, while in the rural south, peasants and intermittent workers were left to the legal charity of the municipalities, as in the liberal era. This chaotic system emerged amid various financial and administrative reorganizations and against the passive resistance of the medici della mutua, the doctors under contract with the sickness funds. As with medici condotti in their relations with communes, doctors were caught in the conundrum of securing a stable, albeit meagre, income from casse mutue or catering to moneyed self-paying clients in a fiercely competitive medical marketplace.

In full continuity with the fascist regime, the first thirty years of the republic nurtured a myriad of casse mutue, which provided unequal protection in terms of eligibility, benefits, and contributions to an increasing proportion of the Italian population. Coverage increased from about 30 per cent of the Italian population in the postwar years to over 90 per cent in 1974, when the casse mutue were officially declared bankrupt. Alternative explanations have been offered for the remarkable continuity of the mutual funds system across different regime types. These include an alleged intrinsic corporative logic of the Italian political system and the enduring impact of the Catholic social doctrine of solidarity and subsidiarity on the Christian Democratic Party, the keystone of all coalition governments of the time. More convincing explanations highlight that the administration of sickness funds through formally autonomous organizations, run by political appointees, shielded governments from taking direct responsibility for their poor functioning and frequent cost overruns, while leaving ample space for creating political machines. Moreover the highly fragmented and differentiated contributory social insurance funds institutionally preserved the privileged position of some occupational sectors and grades of work and emphasized a male breadwinner model of the family.

This scheme provided important political benefits in both regimes. It fitted the fascist politics of political patronage, playing groups off one another, and rewarding with special benefits loyal groups, such as public employees. It also provided important resources for clientelism, as with the practice of extending insurance coverage to middle-class, autonomous workers of great political import such as shopkeepers and small autonomous farmers, without making them fully responsible for financing. The longevity of the mutual funds contributed significantly to health inequalities. Variations in benefits and contributions between and within occupational categories combined with differences in the occupational structure of the industrial north and the agricultural south resulted in a system massively skewed towards the wealthiest, and healthiest, northern part of the country, the Italian version of Tudor Hart's famous "inverse care law" observed in England.

The establishment in 1978 of the *Servizio sanitario nazionale* (SSN) is the watershed event in the organization of healthcare in Italy. The SSN transformed individual entitlements based on occupation and contribution into equal citizenship rights based on health needs. This dramatically changed the logic that had dominated the world of welfare and the processes of social policymaking in Italy during the previous fifty years.

The general principles of the SSN are similar to the British NHS, but there is no documented evidence of a conscious importation of the British model. The Italian season of the movements makes the case for a pivotal role of bottom-up, progressive forces in the birth of the SSN, a point that Giovanni Berlinguer, the founding father of the SSN, was keen to stress in comparing the top-down, political process of the foundation of the NHS. The much-decentralized administration of the SSN, first at the communal and then at the regional level, is a major difference with the original design of the British NHS, where the outcome of the parliamentary battles between Aneurin Bevan and Herbert Morrison favoured centralization. However, in Italy as in Britain, the difficult balance between national institutional and fiscal responsibility, democratic participation, and responsiveness to local needs continues to be a major institutional and political issue.

The institution of the SSN was a dramatic departure from the particularistic orientation of the conservative-continental model of the rest of the Italian welfare system. The institutional and political change put health politics and policy on a different path from the rest of the Italian welfare system. The SSN is a glaring anomaly in the institutional and political history of Italy: an exemplary case of successful modernization of state administration that contrasts with the alleged weaknesses of both civil society and state capacity in Italy, a recurrent theme in its political, economic, and social history. Implementation, however, was a more difficult story. The swift dissolution of the heterogeneous reformist coalition and the oil economic crises of the 1980s hampered the aims of the ambitious plan that remained largely unfinished. The regions faced the challenge of implementing an expansive programme in a context of economic crisis and severe limits on public expenditure and entered in conflict with the strict central government's fiscal policy. The SSN quickly discovered its intrinsic vulnerability to economic crises. Equity of access was one of the main goals of the reform, and the duality of the Italian economy implied that only a national tax-funded system could achieve territorial justice. However, its dependence on tax funding makes the SSN easy prey of fiscal crises since decisions concerning the size of its budget depend on the state of the economy.

The process of implementation proved wanting with respect to the noble principles of the original reform. Patronage and corruption seriously affected the performance of the new SSN, and throughout the 1980s, several schemes were presented to Parliament to "reform the reform". The alleged solution was found in the new political logic of managerialism and competition, which in the 1980s dominated the "epidemic of health reforms" in Western countries, providing an image of apparent "convergence" of their health systems. This started a new political dynamic, which during the fateful decade of the 1990s also changed the face of the SSN. The SSN political history mirrors the British NHS also in its Thatcherite shift "from paternalism to consumerism, from need to demand, from planning to choice" or , in essence "from church to garage" as Rudolf Klein famously put it. Two distinct and opposing waves of reforms substantially changed the organization and functioning of the SSN while preserving its fundamental principles. The 1992 health reform was prompted by two major political and economic crises and legislated strict fiscal policies inspired by New Public Management to get politics out of health. The USLs were transformed into public enterprises (the principle of *aziendalizzazione*) and managed by a director general, a sort of chief executive officer, appointed by the region which was to be responsible for expenditure overruns (the principle of *regionalizzazione*). *Comuni* were stripped of all their administrative responsibilities for running the USLs. Regionalization invoked the standard democratic merits of federalism, in particular the greater proximity of federal governments to the preferences of local citizens. In fact, the main goal was to increase the financial accountability of the regions by obliging them to cover their deficits through increasing own-source taxes or excluding citizens from SSN coverage. To this end, regions were allowed to let their citizens opt out the public system and join private organizations competing with the SSN. This would change the fundamental principles of the SSN and triggered a second reform in 1999, with the explicit purpose of reaffirming the original principles of universalism, comprehensiveness, and public funding that were to be pursued with new policy instruments.

The 1990s opened a new phase in the history of the SSN. The reallocation of responsibility in healthcare galvanized the regions into an intense process of competitive region—building aimed at strengthening their political identity and visibility and subtracting policy space from the central government. Competition among the regions and between them and the central government continued for most of the decade. Initially, the principal issue at hand concerned the choice of the organizational model for the regional health services, but later the question of intergovernmental relations took centre stage. The reform of the Constitution in 2001 marked the high point in the process of "devolution", establishing a new division of power between the state and regions, and again health took centre stage. Regions were given the power to legislate in all areas, except those expressly reserved to the state. In the case of healthcare, the state maintained "exclusive powers" in the definition of the national healthcare entitlement (*livelli essenziali di assistenza*) and the fundamental principles of the SSN. These were not explicitly spelled out, as with the five principles of Canadian Medicare, but had to be deduced from existing legislation. The constitutional reform provoked intense conflict over the allocation of legislative power between the two levels of government. Part of the problem arose from the rudimentary "institutions of federalism" existing in Italy, which are based on the so-called Conferences System. Its opinions (*pareri*) and agreements (*intese*) over *accordi* or broader *patti* are the fundamental acts through which the regions, through their presidents, take part in the national legislative process on matters of common interest with the central government. The product is, however, a sort of the "executive cabinet federalism" lamented in Canada, whose output is soft legislation with uncertain legitimacy under Italian law.

In the last twenty years, the SSN has entered its third phase. Health policymaking has operated in the shadow cast by the towering Italian public debt, by far the largest in the European Union, and by the volatility of the Italian economy. Increased pressure from the European Union to control the growth of public debt explains the calls for containing health expenditure, despite the relatively low public spending of the SSN compared to other OECD countries. The effect of the Europeanization of Italian fiscal policy, a strategy that has been followed regardless of the government in power, has been to focus on the financial aspects of health policy, downgrading questions like health inequalities and service organization, now in the remit of regional governments. Health policies designed at the national level are essentially seen as organizational corollaries to the financial pacts, with periodic re-negotiation of the array of measures necessary to meet global and specific spending ceilings and capacity limits, which then become conditions for receipt of state transfers.

A new sense of impending crisis followed the onset of the Great Recession of 2008. The initial adoption of austerity politics has shifted into a decade of fiscal stringency. Two strategies responded to the severe underfunding of the SSN. The first aimed at consolidating and intensifying fiscal authority of the central government over the regions and their hospitals, now subject to specific, individual recovery plans. The second entailed revamping, streamlining, and integrating primary health and social services to expand chronic care programmes, particularly for the elderly, and at the same time contain their cost. Both strategies were bound to fail, as the first policy was short-lived and the second too far-reaching. Recovery plans were outliving their effectiveness in generating fresh resources from regional taxpayers and service users since most of the regions officially declared "in difficulty" had exited their Piano di rientro. Medical and financial outcomes (if any) of the latter would only show in the very long term. The outbreak of the COVID-19 pandemic changed everything.

While initially enhancing popular trust in the SSN, the COVID-19 pandemic has also dramatically increased its financial difficulties, amidst the economic and social crisis of the country. Current difficulties, which investment funds from the Next Generation EU cannot address, have restored two traditional "solutions" to the ills of the SSN. The first envisages the institution of a second, private financial "pillar" to fund services excluded from SSN coverage. The second entails the financial secession of the wealthiest regions of Lombardy, Veneto, and Emilia-Romagna, which have asked to retain their "tax surpluses", the difference between central taxes and transfers, to fund their own policies. The future of the SSN, as we know it, has never been so uncertain.

Ruptures and Continuities

The very short summary of the course of public health policy in Italy shows that continuity prevails over ruptures, which radically change the logic of the existing system. Major changes to the existing framework are rare events, among which we can identify Crispi's laws of 1888 and 1890, the 1927 Labor Charter, the 1978 institution of the SSN, and the legislation of the 1990s, following the "reform of the reform" of 1992. Moreover, when major changes occurred, their emergence and/or their implementation were usually slow and subject to multiple adjustments. Examining the full "life cycle" of the major episodes of change, understood as a process of emergence, acceptance, and implementation, makes Carolyn Tuohy's classification of policy change in her famous "Remaking health policy" suggestive but ill fitted to the Italian situation. According to Tuohy, given a window of opportunity, four strategies of change are possible, which combine its scale and pace: big bang (large scale, fast pace); blueprint (large scale, slow pace); mosaic (multiple small scale, fast pace); incremental (small scale, slow pace). A few examples from the history of Italian policymaking explain why examining how processes unfold in time is essential and provide grounds for further studies.

During his first spell in government (1887–1891), Crispi's "parliamentary dictatorship" hastened the enactment of sweeping legislation that reorganized the administrative structure of the unitary state and expanded its functions in public health and the organization of public beneficence. At first sight, both health-related laws seem cases of big-bang reform, with large scale at fast pace. The Crispi Public Health Act of 1888 however is also a classic case of slow emergence, as it took twelve years and several complete revisions for the original 1876 Parliamentary Commission report to be passed into law. By contrast, a large parliamentary majority swiftly enacted the 1890 secularization of Opere pie, which however twenty years later was still poorly implemented (see p. 18). Slow emergence does not imply quick acceptance: several outlines of the SSN were drafted in the late 1960s, and casse mutue were abolished in 1974, but it took four more years to discuss and integrate a dozen Parliamentary bills ending with the 1978 reform. Quick acceptance and swift enactment seem enemies of prompt and full implementation: the 1992 "reform of the reform" was enacted in record time and apparently qualified as a big-bang reform but encountered strong resistance, which prompted a cascade of corrective legislation, and staggered its implementation. Both the 1978 and 1992 legislation reflect a polity that requires the negotiation of large-scale reform packages among several competing actors with independent power bases. However, while the compromises that watered down the original legislation of 1992 occurred in the decade after its swift enactment, they were negotiated ex ante during the slow institution of the SSN. Giolittian health and social policies resulted in a framework resembling Tuohy's "mosaic" model of change, the result in part of Giolitti's personal policy style but also of liberal reservations about state intervention. Giolitti never embarked on a politics of comprehensive reform but tinkered with various aspects of existing legislation. However, his innovative health policies provided ad hoc, vertical programmes targeted against the scourges of the time such as pellagra and malaria.

Programmes against specific diseases were clearly embedded in the polity of their time. Malaria, for example, was dealt with in entirely different ways by each of the three italian political regimes. Countering malaria was one of the effects of the fascist "totalitarian" policy of bonifica integrale, which entailed a complex and comprehensive range of hydraulic, agricultural, housing, and human interventions that greatly expanded the reach of earlier liberal projects. In fascist "totalitarian" ideology, integral land reclamation was instrumental to the battaglia del grano to boost cereal production to reach self-sufficiency and serve the imperial aspirations of the regime. The wide reach of the fascist approach to contain malaria is at the opposite ends of the technocratic models of disease-specific interventions developed both in the late liberal era and after the Second World War. At the dawn of the Twentieth century, a series of disease-specific laws designed a programme of *chininizzazione integrale* of the peasants living in designated malarial areas with *chinino di Stato*. The strategy of the early postwar years in Lazio and then in Sardinia was also disease specific and technology driven but aimed at the more ambitious goal of mosquito eradication through environmental sanitation by DDT. Supervised by the Rockefeller Foundation with Alberto Missiroli as the local agent, the Sardinia project was part of the Marshall plan and quickly became intensely politicized in the toxic climate of the early Cold War.

Major policy differences can also be found within the same polity, reflecting only in part the process of knowledge of the disease. The responses to the waves of cholera that hit the country during the liberal period exposed the changing political and social cultures of the various governments of the Destra and Sinistra Storica, adapted to changing scientific knowledge, and tried the capacity of national and local governments, testing the effectiveness of previous interventions. To the first severe epidemic of 1865–1868 in the south, the state engaged in the war against brigandage opposed a strategy of heavy-handed military intervention aimed at controlling social unrests through harsh cordoni sanitari. The 1884–1887 cholera outbreak opened a window of opportunity for significant changes in the intervention of the liberal state. Competing strategies to contain the cholera epidemic opposed traditional "measures of restriction" such as quarantine, to "measures of prevention", such as sanitation and the prompt isolation of cases. The ensuing debate opened the window for a sort of an Italian "revolution in government". New technocratic institutions were created at the central and local level, more attuned to the emerging scientific revolution brought about by the rapidly evolving germ theory of diseases, and a major reform of public health was eventually enacted in 1888. The *legge del colera* o *legge per Napoli* as it was popularly called, provided unprecedented, massive financial support from the state to urban renewal and sanitation, initially of the old city of Naples, but quickly extended to major cities, such as Milan and Turin. The institutional impact of the 1884–1887 epidemics was therefore a prototypical example of cholera as "the good epidemic" disrupting old policies and "forcing" the adoption of apparently unfeasible achievements.

While the cholera epidemic of 1884–1887 led to major transformations in health organization, the two world wars did not open a window of opportunity for significant health policy change in Italy. The national mobilization for the First World War was not the watershed for the development of an interventionist social state observed in other European countries. A plan for compulsory health insurance developed by the *Commissionissima per il Dopoguerra* immediately floundered and was never debated in Parliament. The only social legislation adopted during the First World War was the 1917 extension of accident insurance to agricultural workers, which was twice postponed and implemented only in 1919. After the Second World War, while in most European countries "war and welfare went hand in hand", in Italy the political and economic elites rejected the Beveridgean universalistic approach to social protection adopted in Britain and maintained the Bismarckian corporatist system of social insurance in full continuity with the fascist regime. Even the moderately expansive proposals of the *Commissione per la riforma della previdenza sociale* chaired by Ludovico D'Aragona were quietly dismissed by De Gasperi's government as unaffordable. From then on, economic crises have been the main forces opening windows of opportunity for significant institutional changes, which have usually seen health policy at the forefront. In these events, the medical profession has rarely shown significant political influence.

The Role of Organized Medicine

Organized medicine has always claimed both social and cultural authority on health policy decisions but can hardly be seen as a dominant force in policy making since the early days of unification. The extension of the piedmontese Rattazzi's law and the public health law of 1865 ignored the *Associazione Medica Italiana*'s (AMI) claims of authority in the new health policy of the unitary state based on doctors' participation in the war of independence as well as on its command of professional knowledge. The *Associazione Nazionale Medici Condotti* (ANMC) engaged in a long and unsuccessful fight with municipal authorities for stability and control of working conditions against political and administrative interference. After the long ban of free associations in the fascist regime, the *Federazione Nazionale degli Ordini dei Medici* (FNOM) gained negotiating powers with *casse mutue* in the 1960s but was excluded from the crucial processes in the founding of the SSN, to which some medical organizations instead intensely partecipated (see Chap. 7). This radically differs from the British experience of instituting the NHS, which Aneurin Bevan negotiated with the British Medical Association and the Royal Colleges. However, when the strategy of "political theatre" to reject the law failed, FNOM relied on an effective "quiet politics" aimed at influencing its implementation from behind the scenes. The passing of Bindi's reform brought back a sort of old-style confrontation with organized medicine over the issue of doctors' dual practice, i.e seeing privately paying patients in public hospitals. With the rise of powerful corporate actors such as the pharmaceutical and technology industries, the power of organized medicine has further diminished, also because of its fragmentation in sectorial unions and a plethora of scientific medical associations organizing specialties and subspecialties. It seems fair to say, in summary, that, with the significant exception of HIV/AIDS policy (see page 216) doctors have not played a major role in the politics of health policy formulation. A rather different picture emerges, of course, when looking at policy implementation, which necessarily requires some co-operation from the organized profession.

Population Health

Over the last 160 years, the health of the Italian population has improved enormously. Starting in the late 1880s, economic development, public works, and state interventions that brought to bear medical advances on population health fuelled a significant reduction in mortality and an even greater increase in life expectancy at birth. Initially, the greatest advances were made by preventing diseases from occurring through sanitary measures, vaccination, and social improvements, as with typhoid, smallpox, and tuberculosis. Total mortality fell, particularly in the younger ages, and not only life expectancy, but also the number of years that Italians expected to live in good health increased. Health improvements have occurred at different times and with different speeds across the country, contributing to the historical social and economic divide between the North and the South, still one of the most controversial issues in Italian economics and politics. By the turn of the nineteenth century however, the gap in mortality with the leading European powers of Britain, France, and Germany was still significant and a matter of concern for the medical elite in Parliament. Campaigns against pellagra and malaria, the twin scourges of the time, are both success stories in the social history of population health for quite different reasons. While quininization contained malaria deaths, pellagra simply "cured itself spontaneously...while scientists were still arguing about its causes" (see page 63).

The gap in mortality and life expectancy at birth significantly narrowed by the middle of the twentieth century: life expectancy in Italy is now among the highest in the world. Successful medical treatment has since made an increasing contribution to reducing mortality. Because of increasing survival, total life expectancy has increased much faster than healthy life expectancy, and years of life spent with long-standing diseases increased. This has had significant implications for the patterns of services, financing, and research in an increasingly older population. The outbreak of the COVID-19 pandemic dramatically changed the disease profile into a "double burden" of chronic, long-standing illnesses along with a highly infectious and deadly disease. The ongoing COVID-19 pandemic has significantly influenced both mortality and life expectancy. Although COVID-related deaths concentrate in the elderly population, life expectancy at birth for both genders in 2020 was at 82 years, 1.2 years less than in 2019. This is the heaviest toll of any event in Italy since the Second World War.

Global Health and National Histories

It may seem strange to produce a book narrating a national history of public health policy at a time when the COVID-19 pandemic reminds us that health is a global event. However, the history of health and healthcare in Italy shows that the circulation of goods, people, and ideas has always contributed to the "unification of the globe by disease", to take Le Roy Ladurie's famous turn of phrase. Nations' health, national welfare states, and organized health systems have always been influenced by commerce and cross-border circulation of ideas. Moreover, "unification of the globe" not only was a matter of epistemic communities but influenced domestic as well as foreign policy and entailed a new international diplomacy. Italian government's attempts to curb the regulations established by the Paris International Sanitary Conference governing trans-oceanic circulation of people and goods during the cholera epidemic of 1910–1911 influenced domestic policies to contain the spread of the epidemic and caused diplomatic incidents with Argentina and Uruguay and continuous tensions with the United States. Old protectionist strategies have been resurrected in modern times by the COVID pandemic.

The development of public and private International Health Organizations, from the League of the Nations to the World Health Organization and the Rockefeller and the Gates Foundations, along with the growing interest of the European Union in regulating health-related matters and its expenditure, has intensified the circulation of ideas, people, and, increasingly, financial policies. This has impacted Italy principally in the 1990s, when a process of active learning from international experiences took place (see Chap. 8). However, the alleged "hybridization" of national health systems and their supposed "convergence" have often been limited more to language than organization, as we have seen with the notional import of the purchaser-provider split and quasi-market in the 1990s. The ongoing COVID-19 pandemic has greatly increased the Europeanization of national health policies. Yet the pandemic has also made it more apparent than ever how wide and widening is the Great Divergence of “the West” compared to “the Rest”.

Index[1]

A
Agriculture, 1–4, 10, 20, 29, 40, 42, 52, 100, 101, 115, 123, 143, 145, 149
AIDS/HIV, 212–217, 261, 269
Alto Commissariato per l'Igiene e la Sanità Pubblica (High Commissioner for Hygiene and Public Health, ACIS), 128–130, 147
Amato, Giuliano, 189, 192, 193, 195, 202
Anselmi, Tina, 136, 170
Apulia (Puglia), 2, 4, 6, 20, 36, 38, 44, 45, 51, 98
Ascension speech, 41
Autarchy/autarky, 91, 115
Aziende sanitarie locali (ASLs), 190, 191, 195, 196, 200, 203, 212, 225, 256

B
Baccelli, Guido, 18, 32, 33, 46, 50, 53
Badaloni, Nicola, 42, 46
Barletta, 38
Basaglia, Franco, 169
Battle
  for births, 92, 95
  for wheat, 92, 98
Berlinguer, Giovanni, 53, 146, 148, 159, 166, 168, 170, 177
Bertani, Agostino, 3–4, 11–13, 57
Bevan, Aneurin, 176, 217, 279, 285
Beveridge, William, vi, 127, 129–132, 218
Bindi, Rosy, 198–200, 202, 203, 285
Bizzozero, Giulio, 4
Bonifica/land reclamation, vi, 44, 48, 95–99, 109, 283
Bottai, Giuseppe, 99
British Medical Association (BMA), 176, 285

[1] Note: Page numbers followed by 'n' refer to notes.

F. Taroni, *Health and Healthcare Policy in Italy since 1861*, https://doi.org/10.1007/978-3-030-88731-5

C
Cabrini, Angiolo, 24
Cannizzaro, Stanislao, 18
Carta del Lavoro/Labor Charter, 92, 99–101, 277
Casse mutue, v, 99–101, 103, 106, 107, 109–112, 123, 129, 130, 133, 142–145, 147–149, 152, 158–162, 166, 167, 171, 178, 211, 213, 277, 283, 285
Catholic Church
  cholera Naples, 34
  Congress Milan 1880, 59
  HIV/AIDS infection, 216
  questione romana, 2
Celli, Angelo, 4, 7, 22, 24, 32, 39, 43–47, 49, 53, 96
Celli-Fraentzel, Anna, 45, 62
Chain, Ernest Boris, 127
Cholera, vi, 7, 12, 14, 19, 24, 29, 31–39, 75, 76, 170, 218, 284, 287
Civiltà Cattolica, 16, 34
Cold war, 123, 126, 137, 284
Comuni (communes/municipalities), v, 2, 3, 5, 7, 9, 13, 61–63, 109, 129, 214
Condotte mediche (municipal dispensaries), 9, 12, 48, 49
Condotti, see Medici condotti (municipal doctors)
Congregazioni di carità (congregations of charity), 17–19, 60
Congressi degli Scienziati (Scientists' Meetings), 3, 8
Consiglio, Placido, 78, 79
Consorzi Provinciali Antitubercolari (Antitubercular Provincial Consortia, CPA), 104
Constitution/constitutional court, 16, 46, 121, 127, 128, 133, 134, 136, 142, 143, 165, 167, 171, 173, 176, 191, 196, 200, 204, 205, 231, 232, 234–236, 239, 240, 247, 265, 280
Consultori materno-infantili, 168
Corporazioni, 92, 99
Corradi, Alfonso, 32, 57–59
COVID-19, v, vii, 73, 83, 244, 247–249, 257–269, 275, 281, 287
The creation of the Servizio sanitario nazionale, 165–182
Crispi, Francesco, 3, 12, 14, 17–19, 22, 23, 30, 32, 33, 35, 36, 49, 51, 52, 60, 62, 276, 282
Crispian laws
  public health, 12–14, 30, 32, 50, 52, 276
  Opere pie, v, 2, 14–17, 19, 22, 23, 49, 54–56, 58, 60, 63, 77
Croce, Benedetto, 14, 20
Croce Rossa/Red Cross, 77

D
D'Aragona, Ludovico, 134–137
DDT, vi, 98, 125, 126, 284
De Cristoforis, Malachia, 56
De Gasperi, Alcide, 134, 135, 285
De Lorenzo, Francesco, 182, 197
De Marsanich, Augusto, 131
Depretis, Agostino, 11, 33–35
De Rerum Novarum (of new things–Leo XIII), 18
Devoto, Luigi, 53, 86
Direzione di Sanità, 12, 19, 109
Draghi, Mario, 268

E
Einaudi, Luigi, 131, 132
Emigration, 20, 34, 36, 37, 155

Emilia-Romagna, vii, 4, 40, 167, 196, 224, 248, 257, 261, 263, 282
Enti pubblici/parastate agencies, 107
European Recovery Plan (ERP), 127

F
Federazione Nazionale dell'Ordine dei Medici (National Federation of Doctors Orders, FNOM), 155, 159–161, 175, 176, 203, 285
Flebotomi/phlebotomists, 48, 50
Forlanini, Carlo, 104, 106
Fortunato, Giustino, 39, 45, 46
Franchetti, Leopoldo, 15
Futurism, war, 74, 79, 113

G
Gemelli, Agostino, 78, 79
Gini, Corrado, 79, 94
Giolitti, Giovanni, 17, 19–24, 36, 38, 42, 44, 46, 51, 62, 63, 74, 75, 82, 102, 160, 276, 283
Giornale della (Reale) Società Italiana di Igiene, 6, 37
Giovanardi, Augusto, 128, 129
Golgi, Camillo, 42, 44, 46
Grassi, Giovanni Battista, 44, 46, 96
Guzzanti, Elio, vi, 156, 214

H
Hackett, Lewis Wendell, 44, 97, 98, 109, 115
Health and healthcare in the liberal State, 29–64
Health in the making of a Nation, 1–24
Health under the Fascist State, 91–115
Hospitals
  hierarchical order, 156
  medicalization, 17, 30, 60
  petragnani law, 156, 202
  private patients, 114, 203
Hygiene and beneficence/congress (1880), 56

I
Illiteracy, 1, 7
Industrial accidents, 21, 51, 52
Industrialization, 2, 20, 36, 39, 51, 53
Insurance
  against illness, 112, 145
  against occupational diseases, 53, 99, 100, 108
  against occupational injuries, 21
  against tuberculosis, 92, 99, 100, 102–106, 109, 145, 146, 161, 277
Istituto Nazionale Fascista Assistenza Infortuni sul Lavoro (National Fascist Institute for Insurance in Industrial accidents, INAIL/INFAIL), vi, 108, 147
Istituto Nazionale Fascista della Previdenza Sociale (National (Fascist) Institute of Social Security, INPS/INFPS), 129, 130, 134, 145–147
Istituto Nazionale per l'Assicurazionei Malattia/Istituto Nazionale Fascista Assistenza Malattia (National (Fascist) Institute for Sickness Assistance/Insurance, INAM/INFAM), 130, 135, 142–151, 153, 161, 162, 198
Istituto Superiore di Sanità (National Institute of Health, ISS), 114, 127, 214, 215, 257, 262

J
Jacini, Stefano, 4, 55

K
Koch, Robert, 32, 36, 44, 46, 106

L
Land reclamation (*Bonifica*), vi, 44, 95–99
Lanza, Giovanni, 10
League of Nations (LON), 97, 98
Lega Nazionale contro la Malaria (National League against Malaria), 46
Lister, Joseph/Listerism, 57
Lloyd George, David, 23, 62
Lombardy, 1, 4, 8, 9, 13, 15, 40, 41, 55, 76, 110, 167, 180, 196, 199, 201, 204, 224, 236, 248, 257, 258, 261, 263, 266, 267, 282
Lombroso, Cesare, 39, 41–43, 78, 79
Luzzatti, Luigi, 51, 52

M
Maccacaro, Giulio, 170, 217
Maggiorani, Carlo, 2, 3
Malaria, vi, 2, 4, 7, 22, 24, 29, 30, 43–48, 51, 74–76, 84, 93, 95–99, 121, 122, 124–126, 283, 286
Manson, Patrick, 44
Mantegazza, Paolo, 51
Mariotti, Luigi, 148, 156–158, 166
Marotta, Domenico, 114
Marshall plan, 284
Medici condotti (municipal doctors), v, 8, 9, 11–13, 22, 42, 43, 45, 49–51, 58, 60–62, 82, 85, 101, 110, 111, 113, 114, 160, 276
Medici della mutua, 112, 155, 159, 277
Medico provinciale, 13, 109
Messedaglia, Luigi, 43
Ministry of Health, 81, 107, 127, 128, 146–148, 157, 160, 178, 199, 221, 244, 261
  Ministry of "salvo," 157
Ministry of the Interior, 10, 12, 33, 109, 128, 148, 152
Missiroli, Alberto, 121, 124, 125, 284
Monti, Mario, 247, 249
Morelli, Eugenio, 103, 104, 112
Moro, Aldo, 166
Movements, season of
  mental health, 256
  women, 165, 168
  workers' model, 168
Mussolini, Benito, 20, 41, 75, 91–97, 99, 104, 108, 112, 114, 122, 123

N
Naples
  cholera, 32, 34, 36, 37, 170
  law for Naples, 35
National Insurance Act (GB, 1911), 106
New Man (Uomo Nuovo), 93, 94
Newsholme, Arthur, 85
Next Generation Plan EU, 268, 282
Nitti, Francesco Saverio, 15, 16, 35, 43, 82
Nuova Antologia, 33, 86

O
Opera Nazionale Combattenti (National Organization of Combatants, ONC), 96
Opera Nazionale Maternità e Infanzia (National Organization for

Mothers and Children, ONMI), 94, 108, 109, 129, 168
Opere pie (pious works; charities), v, 2, 14–19, 22, 23, 49, 54–56, 58, 60, 63, 77, 202, 276, 283
Ordine dei medici (Order of Doctors), 22, 62, 113

P
Pagliani, Luigi, 12, 14, 19, 32, 33, 44, 50, 57
Panizza, Bernardino, 13
Peasants/peasantry, vi, 3, 4, 21, 29, 39, 41–43, 45, 75, 96, 98, 100, 277, 283
Pellagra, vi, 2, 4, 7, 22, 24, 29, 39–43, 51, 77, 283, 286
Pende, Nicola, 94, 96, 114
Penicillin, Fabbrica, 126, 149
Petragnani, Giovanni, 114, 152, 156, 202
Piani di Rientro, 240–244, 262
Piedmontization, 8–10
Pini, Gaetano, 56
Pontine Marshes, 4, 51, 96–98, 109, 124
Postwar roads not taken, 121–137
Po Valley/Po River, 1, 42, 44, 45, 51
Provinces (province), 8, 44, 54, 75, 148, 156, 160, 193, 238, 239, 261
Psychiatry
  Basaglia, 169
  fascism, 113
  World War One, 113

Q
Questione meridionale/Southern problem, 2
Quinine, 43–48, 76, 84, 97, 98

R
Racial Laws, 113
Raseri, Enrico, 32, 48, 54, 58, 59
Rattazzi, Urbano, 8, 9, 285
Regions, 6, 32, 76, 125, 151, 166, 187, 211, 232, 247, 277
Resistenza/resistance, 122
Ricasoli, Bettino, 10
The rise and fall of the mutual jungle, 141–162
Rockefeller Foundation, 97, 98, 102, 115, 125, 126, 147, 284
Ross, Ronald, 44, 45
Rossi Doria, Tullio, 36, 95
Rudini, Antonio di, 19, 20, 36

S
Salvemini, Gaetano, 20, 74
Sanarelli, Giuseppe, 22, 42, 43, 49
Sanatoria, 96, 102–107, 109, 111, 113, 151, 277
Sanitary Conferences, 31, 34, 287
Sardinia/Sardegna, 4, 20, 44, 49, 98, 122, 125, 126, 283, 284
School of Public Health, 12, 50
Serao, Matilde, 4, 34
Serpieri, Arrigo, 95, 96
Sicily/Sicilia, 2, 4, 7, 15, 17, 34, 44, 49, 122, 180, 224, 241
Socialism, Socialist Party, 14, 18, 45, 46, 51, 52, 75, 79
Society for the Study of Malaria (Società per gli studi della Malaria), 45
Sonnino, Sidney, 18, 45
Soper, Fred, 125
Sormani, Giuseppe, 3, 41, 51
Spanish flu, 73–86
Statistics, mortality hospital, 3, 31, 40, 41, 58, 98, 212
Strambio, Gaetano, 9, 49, 56
Swellengrebel, Nicholas, 97

T
Tangentopoli, 187
Tommaso-Crudeli, Corrado, 18
Torelli, Luigi, 43, 44
Tropeano, Giuseppe, 36, 37, 46
Tuberculosis, vi, 2, 6, 7, 24, 30–32, 74–76, 81, 82, 93, 100, 102–107, 111, 113, 121, 124, 146, 148, 286
Turati, Filippo, 99
Typhus/typhoid, 5, 6, 31, 76, 121, 124, 218, 286

U
Ufficiali sanitari, 13, 22, 23
Unità sanitarie locali (USL), 172–175, 177–181, 188, 190, 191, 195, 205, 215, 233, 280
United Nation Relief and Recovery Agency (UNRRA), 123–126
Uomo Nuovo/New Man, 93, 94
Urbanization, 20, 93, 94

V
Veneto, 4, 6, 13, 40, 41, 84, 85, 96, 167, 180, 248, 257, 261, 263, 267, 282
Verbicaro, 38, 39
Villari, Pasquale, 4, 15, 18

W
Welfare, v, vi, 21–23, 80, 82, 83, 92, 107, 108, 121, 122, 127, 128, 130, 143, 165, 177–179, 211, 212, 219, 220, 231, 240, 278, 279, 285, 287
particularistic, 21, 211, 279
White Mario, Jessie, 4, 11, 34
Women/feminism, 17, 20, 21, 44, 45, 50, 62, 74, 76, 93, 94, 112, 165, 167, 168, 276
World Health Organization (WHO), 126, 127, 133, 169, 212, 222, 223, 251, 287
World War I/Great War, 4, 19, 29–64, 73–86, 93, 95, 96, 102, 124, 128, 284
World War II, 91, 121, 141, 258, 277, 283–285, 287

Z
Zucchi, Carlo, 33, 57, 59, 60

CPSIA information can be obtained
at www.ICGtesting.com
Printed in the USA
LVHW032042141221
706151LV00001B/70